STARS IN YOUR HAND

STARS IN YOUR HAND

A GUIDE TO 3D PRINTING THE COSMOS

KIMBERLY ARCAND AND MEGAN WATZKE

THE MIT PRESS
CAMBRIDGE, MASSACHUSETTS
LONDON, ENGLAND

The MIT Press would like to thank the anonymous peer reviewers who provided comments on drafts of this book. The generous work of academic experts is essential for establishing the authority and quality of our publications. We acknowledge with gratitude the contributions of these otherwise uncredited readers.

This book was set in ITC Stone Serif Std and Avenir LT Std by The MIT Press.
Printed and bound in the United States of America.

Library of Congress Cataloging-in-Publication Data

Names: Arcand, Kimberly, 1975- author. | Watzke, Megan K., author.
Title: Stars in your hand : a guide to 3D printing the cosmos / Kimberly Arcand and Megan Watzke.
Description: Cambridge, Massachusetts : The MIT Press, [2022] | Includes index.
Identifiers: LCCN 2021046827 | ISBN 9780262544153 (paperback)
Subjects: LCSH: Astronomy--Technological innovations. | Three-dimensional printing.
Classification: LCC QB47 .A68 2022 | DDC 522--dc23/eng/20211130
LC record available at https://lccn.loc.gov/2021046827

10 9 8 7 6 5 4 3 2 1

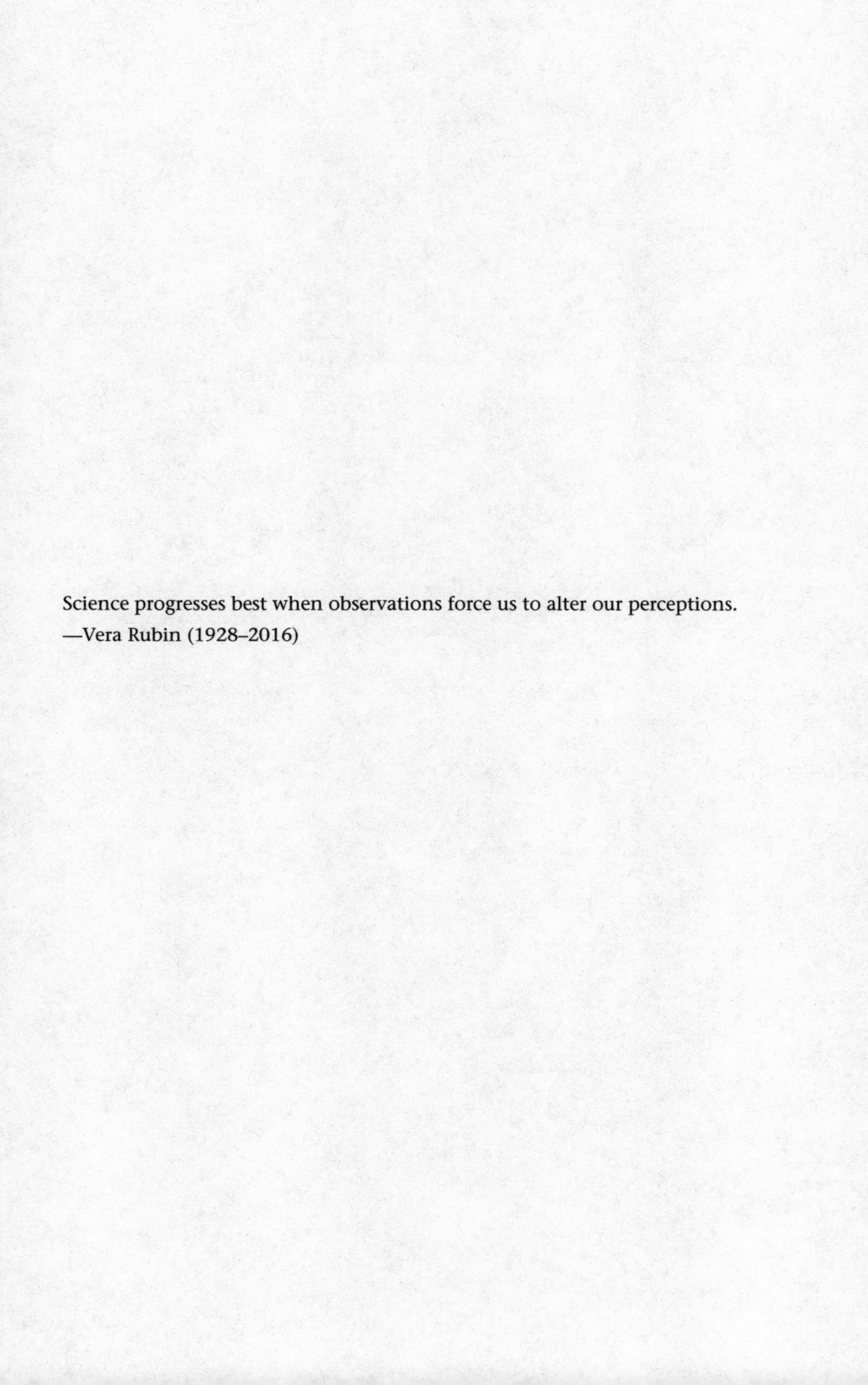

Science progresses best when observations force us to alter our perceptions.
—Vera Rubin (1928–2016)

CONTENTS

1

INTRODUCTION

A mention of *new dimensions* and *space* perhaps invokes an episode from a science fiction show or plot from a futuristic fantasy novel. But the information in this book is not about warp speeds or wormholes. Everything presented in this book is based on research happening here and now and rooted in scientific data.

Astronomers have done remarkably well learning about our Universe, despite having been largely confined to this planet. Of the billions of people that have lived on Earth, only a few hundred have traveled to what is essentially a hair width's distance (a few hundred miles) above its surface. The Apollo program sent a handful of people to the Moon, but our plans to return there and go on to Mars remain unfulfilled, and destinations beyond remain aspirational.

While much has been learned by sending machines into space, these robotic explorers have ventured into only a small fraction of our Solar System. Our planetary neighborhood—which consists of the Sun, the planets, moons, asteroids, and more—is one of thousands that we now know about within our Milky Way galaxy. Our Galaxy is but one of billions of these cosmic island universes, to borrow a phrase from the famous astronomer Edwin Hubble, each hosting billions of stars and a sea of planets.

The ingenuity and creativity needed for these achievements is indeed inspiring, but astronomers have consistently run into the same obstacle: looking at the flat projection, or two-dimensional (2D) view, the sky has offered them.

From Ptolemy to Abd al-Rahman al-Sufi to Vera Rubin, every astronomer across the centuries could only see data of objects in space from a certain point of view. A paleontologist can pick up a discovered fossil and look at it from all sides, and a botanist can examine the backside of a leaf under a microscope. Astronomers have never been able to do such things.

Of course, modern astronomers have an amazing suite of telescopes and observatories on the ground and in space that the scientists who went before them would consider revolutionary. Today, scientists can study objects in space in every type of light—spanning from radio waves to gamma rays—and beyond into gravitational waves and neutrino signals.

Yet still, today's astronomers have generally faced the same hurdle as their scientific foremothers and forefathers: they could not go beyond the two-dimensional window our view of the Universe presented. Until now.

Astronomers have begun to take the vast reservoirs of data available to them and search for ways to go beyond a flattened view of space. Like learning a hidden language that was unknown to them previously, astronomers are now deciphering the code found in these data that reveal the three-dimensionality of space.

The *golden age of astronomy*, as some in the field call our current era, has arrived nearly simultaneously with giant leaps in technology. Supercomputers have continued their march to ever greater computational power as each year passes. What was seen as extraordinary in computing just a short time ago has become mundane. The computing chops of the current generation of smartphones, as an example, could only be found in special facilities not long ago.

Like two streams joining to create a mightier river, the separate advances both in telescopic capabilities and computer technologies have created a new way to explore the Universe: in three dimensions.

The possibilities of 3D technology (including 3D imaging, 3D printing, virtual reality, and holograms, for example) are exploding. Today, researchers can use 3D imaging to map the inside of the human brain or explore the structure of a newly emerging virus. We can now physically construct everything from sections of bridges to automotive parts to prosthetics using 3D printing.

But why try to learn about the Universe in 3D? First, it benefits science and the scientific process. Astronomers who think they understand the geometry and physics of objects by looking at traditional 2D data can discover new elements or aspects in the data that they didn't expect. There is new science to be learned *only* by going into the third dimension.

The second reason is that these 3D models and the resulting 3D prints are tools to engage with people in innovative ways. Many of us have seen stunning images produced by the world's telescopes, but it's another experience to fly virtually through one of these vistas or hold representations of these objects in your hand.

In short, the 3D Universe helps us go further in our exploration of space while simultaneously bringing space right back to us. It can be a bridge between what is "out there" and what we can experience here. And it gives us the possibility to explore these celestial and cosmic objects in a hands-on, tactile way if we have access to a 3D printer. This opens the door for people who are visually impaired and low-vision, and for any of us who benefit in learning from using the sense of touch.

Most humans—at least in this generation and likely those that immediately follow—will not travel into space. But science and technology have advanced where we can have pieces of space come down to us. It's a new way to teach, learn, explore, and be awestruck by the wonders that our Universe has to offer.

Here we are not solely talking about professional astronomers who can direct the world's most powerful telescopes around the globe and in space. The 3D Universe in this book is available to anyone who is interested in astronomy, 3D technology, our place in the Universe, or just learning about something new. We have written this book in a way that we hope makes the science and the technology it draws on understandable and accessible—with the only requirement being curiosity.

The possibilities of the 3D Universe can be right at your fingertips, no further away than turning the pages of this book. Links to all of the 3D models in this book, instructions on how to use a 3D printer, and other resources and updates are available at the companion website: https://starsinyourhand.pubpub.org/.

2

COLLECTING LIGHT

Astronomers sometimes call telescopes "light buckets." This rather inglorious term refers to one of the main purposes of any telescope: to collect as much light as possible from a cosmic source. Telescopes have become more complex as they have spread into space and across the entire electromagnetic spectrum (a.k.a., the full range of light). However, keep in mind that the name of the game generally is capturing light, sometimes photon by photon.

For many centuries, this collection process was limited to the diameter of the human eye. Once the telescope was invented in the mid-seventeenth century, the possibilities grew. Successive generations of sky-watchers expanded their ability to see fainter and farther objects in space by building bigger and better mirrors and lenses. These increasingly large pieces of glass mimicked what the human eye could do in terms of collecting light, but on a much bigger scale.

The advent of photography and photographic plates led to a new leap forward in the ability to record light. Now astronomers could open their telescopes and control the exposure time (that is, the amount of time light could be received) for as long as the rotation of the Earth and the passing of the weather allowed.

Meanwhile, another revolution for astronomy was taking place as a succession of new windows opened "other" types of light. Beginning in the late eighteenth century and continuing through the early twentieth, scientists learned of the true nature of light that extends much wider than anyone knew across a largely invisible rainbow. As the new types of light were discovered, scientists learned that each had different properties that made them valuable for both applications here on Earth and the information they unlocked about the Universe.

Since the electromagnetic spectrum, or light, is a continuous spectrum based on the wavelength of light (see figure 2.1), it can be sliced and classified in countless ways, and many scientific and engineering fields do so for their own purposes. Here we cover the basic set of categories generally used by astronomers and astrophysicists, and some of the contributions of these types of light to our exploration of the Universe.

Radio waves At the lowest end of the energy range of light, radio waves' long wavelengths (some of which can span a full mile) allow them to pass through many types of material including air and water. Radio waves do not interfere with visible light, the dominant type of light from the Sun, which is why these telescopes can operate during the day. In space, objects like clouds of cool gas, stars, and galaxies are strong radio wave emitters.

Microwaves The next step on the light ladder are microwaves. They have just slightly shorter wavelengths than radio waves, and certain bands of microwaves interact strongly with molecules of oxygen and water (which is why microwave ovens work so well heating up food that usually contains water.) From the imprint left over from the Big Bang, known as the cosmic microwave background, to our Sun, many cosmic objects give off microwaves.

Infrared light Often associated with heat, infrared light encompasses a wide range of wavelengths and is often separated into subcategories. Some bands of infrared reach the Earth's surface, but others do not. Therefore, some infrared telescopes are located on the ground, and some are in space. From exoplanets (planets around stars other than our Sun) to distant regions where stars are born, infrared light shines from sources across space.

Visible light This is the band of light that most people are familiar with. It encompasses the wavelengths that the human eye can detect, it makes up the colors of the rainbow, and it's the band in which our Sun gives

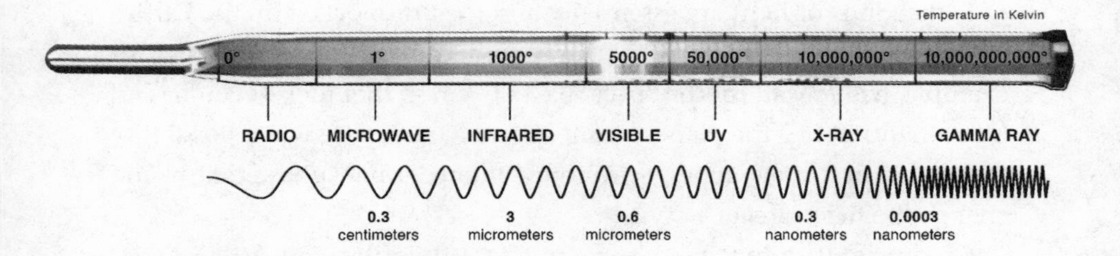

Figure 2.1
The electromagnetic spectrum makes up different kinds of light from radio waves to gamma rays. The wavelength of radiation produced by an object is typically related to its temperature as portrayed in this illustration.

off most of its light. For centuries, visible light was the only type of light that humans could use to study the cosmos as discussed previously. Today we know it represents only a small sliver of the electromagnetic spectrum.

Ultraviolet light With slightly shorter wavelengths than visible light, ultraviolet (UV) is more energetic and thus has the potential to strip atoms and molecules of their electrons. This means it can be both harmful and beneficial to life on Earth, including humans. Looking at space through UV lenses reveals young stars, planets, galaxies, and many more objects that glow brightly in this type of light.

X-rays Best known for their use in dentist and doctor offices here on Earth, X-rays play a crucial role in the investigation of space. X-rays are usually produced by objects that have very high temperatures (upward of tens of millions of degrees). This includes material swirling around or being shot out of a black hole and the debris from an exploded star.

Gamma rays Gamma rays reside on the opposite end of the spectrum of light from radio waves and are the most energetic form of light we know. Gamma rays are typically created by extreme events involving nuclear reactions or particle acceleration and can reveal information about cosmic rays and stellar explosions. Scientists also think a burst of gamma rays signals the birth of a black hole.

Astronomers soon realized that space and the objects that inhabit it tell their true stories across the full span of this electromagnetic spectrum. Each type of light requires its own technology to detect it, which led to the development of new types of telescopes and instrumentation.

Some kinds of light are essentially absorbed or blocked by the Earth's atmosphere, which meant that X-ray and gamma ray astronomy, for example, had to wait for the Space Age to begin so that rockets could send instruments above the planet's atmosphere. Once space was a possibility as a platform for telescopes, astronomy entered a new phase that would change the field forever.

Of course, it didn't happen overnight. While the Very Large Array, a collection of radio telescopes in the high desert of New Mexico, was being constructed in the 1970s, scientists were formulating ideas to put sophisticated telescopes into space. It took decades of planning and construction but beginning in 1990 with the launch of the Hubble Space Telescope, NASA put its "Great Observatories" plan (which also included the Compton Gamma Ray Observatory, the Chandra X-ray Observatory, and the Spitzer Space Telescope) into action. These four telescopes each looked at different types of light far above the distorting and absorbing effects of the Earth's atmosphere.

The turn of the millennium kept this trend of astronomical exploration going across all types of light. Space agencies like the European Space Agency (ESA), the Japan Aerospace Exploration Agency (JAXA), the China National Space Administration (CNSA), the Indian Space Research Organization (ISRO), and others have made astronomy a global effort.

Astronomers collaborate to plan, design, and build new generations of telescopes that will complement one another. Today, there are scores of different telescopes and observatories in various orbits around our planet pointing their gaze outward across space.

While some astronomers have kept their focus on space as the best environment for their research, there is still much to do on the ground. Space offers some very clear benefits, namely escaping the distorting and blocking effects of our planet's atmosphere. There are, however, drawbacks to leaving the Earth's surface. On the one hand, scientists are limited by the size of what can be launched—and the related expense. On the other hand, facilities on the ground can feature "light buckets" as big as engineering, budgets, and imaginations allow.

For decades, the Hale 200-inch telescope at Palomar Observatory in southern California reigned as the world's largest telescope. Hemmed in by how to make a single piece of glass any bigger that didn't warp, buckle, or crack, astronomers began searching for new techniques. By the 1980s, scientists and engineers were working on segmented mirrors, where instead of one solid piece of glass, a mirror is made of connecting individual smaller cells.

In this decade, astronomers are working toward mega-projects that will dwarf the size of previous visible light telescopes. Now under construction in the high deserts of Chile, the Extremely Large Telescope will have a main mirror with a diameter of 39 meters, approximately 127 feet. That's wider than a basketball court end to end.

Meanwhile, planetary scientists were also stretching their reach—literally—out into the Solar System. After the Apollo missions brought samples of lunar soil and rock back to Earth, some scientists turned their attention to Mars and beyond. In 1976, NASA's Viking program became the first American mission to successfully land a spacecraft on Mars and send images back of its surface.

Since then, the space agencies from around the world have sent probes and spacecraft to everything in the Solar System from the Sun to Pluto and many places in between, and even beyond into interstellar space. These robotic explorers are technological marvels in themselves, and they have provided data that would simply be impossible to gather from Earth.

Some planetary explorers have completed orbits around a rocky body, allowing us to have dramatically detailed views. The far side of Mercury or the Moon, as examples, never rotate in our direction, rendering our planet's perspective of these other worlds incomplete. However, we can now construct a complete three-dimensional view of these worlds thanks to intrepid robotic explorers including NASA's Messenger and Lunar Reconnaissance Orbiter spacecrafts respectively (just to name two) that are able to completely circle these bodies. And NASA has announced plans to send two new missions to Venus, one of which will map nearly its entire surface in 3D.

In recent years, scientists have also witnessed the success of a new way to explore the Universe: through gravitational waves. Once considered a mere thought experiment by Albert Einstein, gravitational waves are ripples in the fabric of space-time caused by a violent and energetic event. These tiny perturbations are very difficult to detect, but the Laser Interferometer Gravitational-Wave Observatory, or LIGO, did just that for the first time only a handful of years ago. These twin observatories, located in Louisiana and Washington state, are extremely sensitive triangular laser-based instruments, with each point spaced out four kilometers apart.

The first confirmed detection from LIGO was announced in early 2016. The event was the collision between two black holes about 1.3 billion light-years from Earth. In the years since then, scientists have detected other collisions involving neutron stars (the dense cores remaining from an exploded star) and black holes. In addition to LIGO, which is undergoing improvements, arrays in Japan and Europe have since come online. Combining LIGO's capabilities with Virgo, another triangular gravitational-wave interferometer in the Italian countryside, and the Kamioka Gravitational Wave Detector (KAGRA), deep inside a mine in Japan, will unleash even more new science in the years to come.

Next, we will see what happened when astronomers, computer scientists, and others began to collaborate to take the data captured by these amazing astronomical advances in a new direction.

3

HOW WE SEE THE UNIVERSE IN 3D

In the summer of 2020, a group of astronomers announced that they had discovered a new structure. This was no ordinary object—it was a wall of thousands of galaxies that stretched for at least 700 million light-years across space. (For context, one light-year is equal to about six trillion miles; our entire Milky Way galaxy is about 100,000 light-years across.)

The astronomers dubbed the gargantuan structure the "South Pole Wall." It allows scientists to help piece together our Universe across the largest scales. While scientists once thought that galaxies would be strewn across the Universe in an essentially uniform fashion, we now know that galaxies are laid out in giant arcs and streamers. In between, there are vast gaps and voids where the Universe appears to be a relative cosmic desert.

The South Pole Wall is, in many ways, the scientific descendant of research that began in the mid-1980s. During this time, astronomers Margaret Geller and John Huchra measured the distances from Earth to thousands of galaxies in a slice of the sky using a telescope in southern Arizona. Geller and Huchra plotted the galaxies in a computer program and, importantly, mapped their data in three dimensions—which led to quite a surprise. They didn't find a smooth, evenly distributed sprinkling of galaxies throughout the cosmos as many expected. Instead, they discovered that galaxies lay along thin filaments surrounding giant voids of nothing, like soap bubbles in a sink.

In the years since Geller and Huchra's landmark discovery, other astronomers have gathered more data on different parts of the cosmos, expanding the coverage of these cosmic maps to greater distances and revealing new details within. Each new effort has added to our understanding of the large-scale structure of the Universe. For example, the Sloan Digital Sky Survey, with its automated telescope in New Mexico, has captured wide-angle observations of slices of the sky. One of the primary goals of the project, which began collecting data in 2000 and continues today, is to construct 3D maps of the Universe that build on Geller and Huchra's work.

New projects—either in operation or under construction—that will capture a staggering amount of data points are assuming the mantle of exploring the Universe on some even larger scales in 3D. Consider, for example, the European Space Agency's Gaia observatory, which is currently in space measuring the positions, distances, and motions of upward of a billion astronomical objects. One primary goal of the Gaia mission is to create a 3D map of the multitude of objects throughout our Milky Way and determine how they move. Astronomical software has already been built to take advantage of the Gaia data, providing 3D visualizations showcasing Gaia's 3D map of the sky that we can tour on computers and mobile devices (see figure 3.1).

In a few years, construction of the Vera C. Rubin Observatory (previously known as the Large Synoptic Survey Telescope) should be complete in the Atacama Desert in the high plains of Chile. When it becomes operational, the Rubin Observatory will capture twenty terabytes of images and data *per day* (compare this with the Hubble Space Telescope, which collects about two and a half gigabytes of data per day). The Rubin Observatory will operate the biggest digital camera ever built to capture information of the visible Universe. It will gather data of our entire Southern Hemisphere's sky (covering an area more than forty full Moons combined) every three nights to help scientists analyze and understand the 3D nature of our place in the cosmos.

But there's even more that we can learn about space in 3D right now. For example, on January 1, 2009, the journal *Nature* published a paper online in an unusual interactive PDF format. For the first time in its storied 150-year history, *Nature* let its readers manipulate a cosmic 3D object

Figure 3.1
Gaia's view of the Milky Way as observed between July 2014 and May 2016 is shown in this captivating image. The Gaia observatory from the European Space Agency has been measuring the positions, distances, and motions of about a billion astronomical objects to help create a 3D map of the Milky Way's objects and determine how they move.

in a scientific paper directly, by spinning it around or selecting certain elements to view (see figure 3.2).

With Harvard astronomy professor Alyssa Goodman as lead author, the paper explored the role that gravity plays in molecular clouds. The story of this result, however, began several years before the paper's publication. Goodman, who helped establish Harvard's Initiative in Innovative Computing, was working with a discipline-crossing team of medical researchers, engineers, computer scientists, astronomers, and other experts. This project, which became known as the Astronomical Medicine (AstroMed) Project, fused technology designed for medical imaging—including making 3D images of the brain—and astronomical research through new software.

How do you go from brain scans to black holes? Light, no matter what kind it is (radio waves, infrared, X-rays, etc.) or where it comes from (terrestrial or cosmic), can be split apart. Newton famously discovered this when he used a prism to separate sunlight into the colors of the rainbow. Within this continuum of color, astronomers can identify "spectral lines" of specific atoms and molecules. Spectral lines are caused either by an atom absorbing or releasing a photon. These "absorption" and

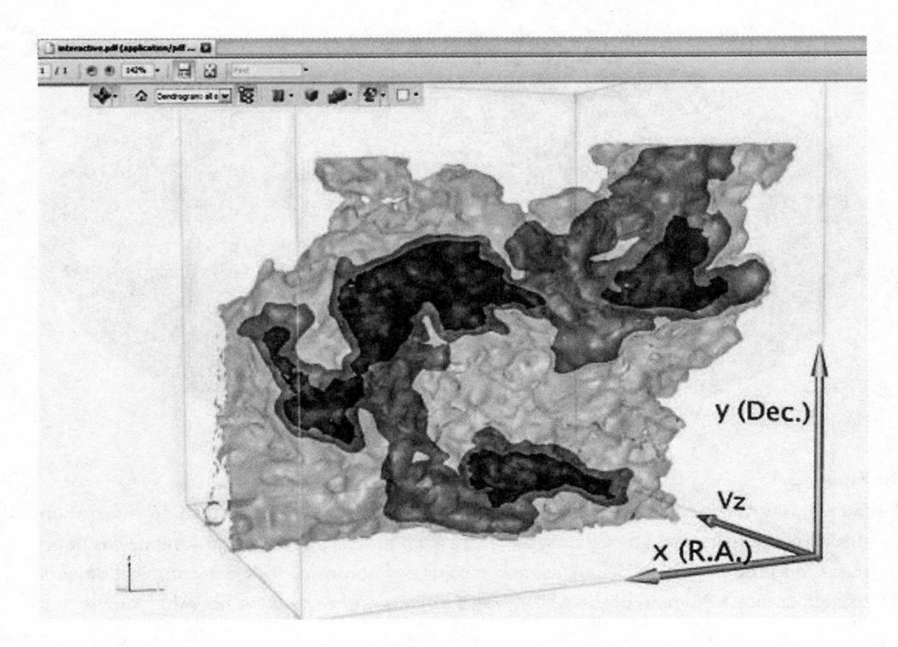

Figure 3.2
This 3D figure of molecular clouds is a screenshot from an interactive PDF of the paper by Dr. Alyssa Goodman and coauthors in *Nature*. Manipulating dimensional information, through rotating the object or turning individual layers on or off, provides a unique way to study the peer-reviewed scientific results of colleagues.

"emission" lines are at specific wavelengths of light, depending on the exact configuration of protons, electrons, and neutrons—akin to the fingerprint of an individual human.

Astronomers can identify specific atoms in the light from a cosmic object and look to see if the spectral lines have shifted at all. Not only is this possible for the rainbow of light we see with our eyes, but also for the spectral lines found in all types of light. Scientists call this change of wavelength, or frequency, the Doppler shift—and it's found in more than just astronomy.

A common example of the Doppler shift is the siren of an approaching ambulance. As the ambulance gets closer to the listener, the volume of the siren increases. The sound of the siren changes in another significant way. The listener will hear a higher frequency for an approaching siren and a lower frequency for one travelling away. Scientists can use Doppler information to measure, for example, moving pressure fronts and precipitation in weather, or blood flow through blood vessels.

Astronomers can measure the shift in light from radio waves to X-rays and determine the velocity of cosmic objects and regions as well as their direction. Since the Universe is expanding, most objects are "redshifted" and moving away from us. The larger the redshift, the farther away an object is. In contrast, something that is moving toward us is being "blueshifted."

However, within individual cosmic objects, the situation can be much more complex and different parts can be moving in different directions (such as stars in a galaxy or galaxies in a galaxy cluster). The amount of shift of the spectral line—that is, how far it is moved from its original position—also gives information about how quickly that space chunk is moving.

While scientists have been using Doppler information provided by spectral lines for over a century for things like distance measurements, astronomers realized they can also use this velocity information to turn their datasets into 3D objects. By exporting the position and velocity data into computer programs, scientists and others could begin to look at data in new ways as specific objects just as they had with the thousands of galaxies across much larger scales.

In 2010, a group used data from several different telescopes to create the first 3D rendering with observational data of the supernova remnant Cassiopeia A, using the Astronomical Medicine software. Supernovas are some of the brightest explosions in our Universe (sometimes briefly outshining entire galaxies), and the debris fields they leave behind contain the chemical seeds for the next generation of stars and planets. The Cassiopeia A 3D model, which incorporated three datasets from optical, infrared, and X-ray light, allowed researchers to explore the remnant from the explosion—and the elements it distributed into interstellar space—as never before was possible.

By 2015, astronomers were beginning to use 3D techniques and observational data to move from studying stellar death to stellar birth. Using data from the Hubble Space Telescope and other observatories, researchers created a 3D model of a portion of the stellar nursery in Messier 16, also known as the Eagle Nebula. This nebula contains one of the most famous objects in astronomy, which has been nicknamed the "Pillars of

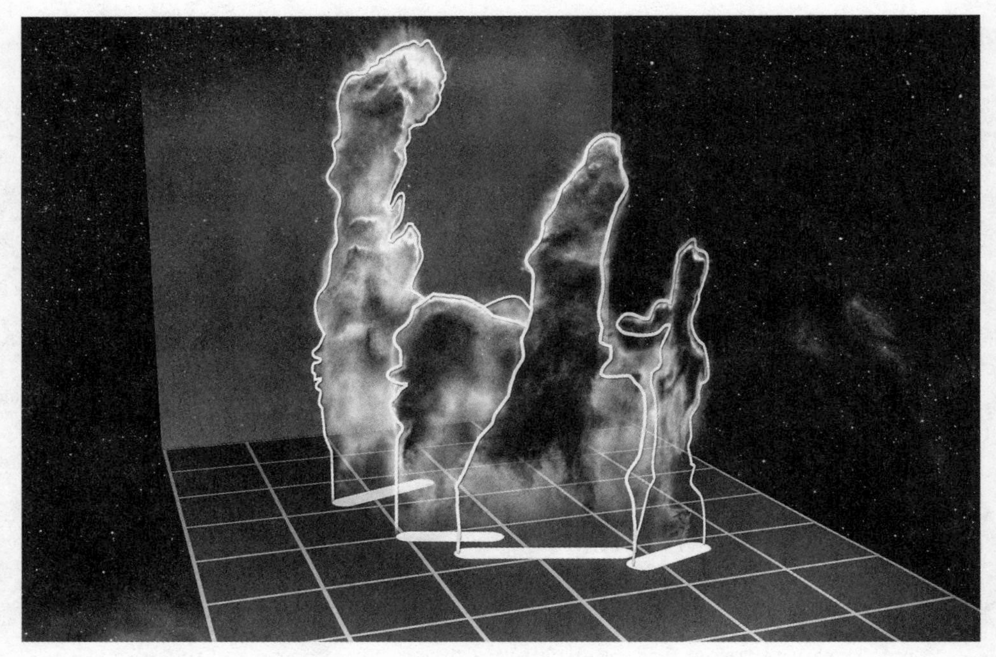

Figure 3.3
The "Pillars of Creation," part of the Messier 16 region, are shown in a 3D space in this image to help demonstrate how the pillars are separated and oriented from each other. From our far-away perspective here on Earth, it is challenging to be able to characterize the relationship between these structures without such 3D information.

Creation" (see figure 3.3) because of its majestic towering structures. The 3D model showed how the pillars are separated and oriented from each other in space—an important achievement that would have been much more difficult using only two-dimensional studies.

Astronomers collect the data in wavelengths across the electromagnetic spectrum (as discussed in chapter 2) to gather clues to many different physical phenomena of cosmic objects including, but not limited to, geometry, temperature, and pressure. Having many kinds of light available—and information going beyond light as well (see for example, gravitational waves in chapter 8)—gives researchers more tools to work with when analyzing such data, whether in 2D or 3D.

Scientists can then also use supercomputer simulations to "reverse engineer," so to speak, the phenomena they observe in the data. By building simulations to match the data delivered by the telescopes, they can manipulate the variables to come into agreement with what they see. If they can get a good fit, then they can have a high degree of confidence that the physics inputted into the simulation represents the same—or at least very similar—physics that is driving the action in real space. Astronomers have done this to build additional 3D models of cosmic objects ranging from stars to galaxies to black holes (examples of which are found in chapters 7 and 8).

Some of the earliest "computers" in astronomy were people, specifically women, who worked largely unrecognized in the late nineteenth and early twentieth century to make measurements by hand and eye that helped make some of the major discoveries of the day possible. The progress of computation abilities once machines were introduced underwent exponential growth. Our modern supercomputers that generate these 3D models are orders of magnitude faster and more powerful than even those that were state-of-the-art just a handful of years ago.

Today's supercomputers run on more than a million processing cores and can handle extremely large datasets, processing them at speeds up to 20 quadrillion (or 20,000,000,000,000,000) operations per second. Such computers help create simulation models of real-world objects or phenomena from real-time weather predictions to visualizing the neural connections of the human brain at scale. In astronomy, simulations can create fluid or hydrodynamic simulations to model events in the Universe based on the data of things like distant galaxies or nearby black holes.

For example, running a mathematical simulation of the center of the Milky Way, constrained by data from telescopic observations on NASA's High-End Computing (HEC) supercomputer located at NASA's Ames Research Center, can take days to compute and render. Ultimately, such a rendering then creates an output that can be translated into a fully immersive 3D view of the massive stars, interactions, and hot gas immediately surrounding the supermassive black hole, named Sagittarius A* (Sgr A*), at the center of our Galaxy.

The raw output of these data-matched simulations can sometimes be challenging to interpret for nonexperts in the field. Therefore, these simulations can also be enhanced for a wider audience by putting them through visualization software where color, texture, and camera angles are improved. Since the early 2000s, 3D programs have migrated from largely the domain of Hollywood studios and professional animators to more readily available off-the-shelf software that many can learn to use. Today, scientists and enthusiasts can leverage various applications to work in such visualizations, from Maya and 3DStudio to Blender and Houdini, and many others.

Now that we know the basics of how we can create these 3D models of objects in space, the question to ask next is: What do you do with them?

4

PUTTING THE 3D UNIVERSE TO USE

Four years before Steve Jobs introduced the first Macintosh personal computer and Apple became a household brand, 3D printing had already gotten its start. In May 1980, Hideo Kodama in Japan patented a technique using ultraviolet light to harden layers of a type of substance, called a polymer, into a solid body.

Then, in the mid-1980s, a process that used lasers to link molecules together, an important underlying feature of additive manufacturing, was invented. By the end of the decade, these advancements would enable development of the basic technology of 3D printing: a machine that melts a filament and deposits it onto a base that can be built up layer by layer until it forms a 3D object.

Fast forward some thirty years and 3D printing can be found everywhere from libraries, hospitals, and schools to the International Space Station (see figure 4.1). Despite its decades-long existence, 3D printing is, in some ways, still in an adolescent phase. Much like the internet in an earlier epoch, many people are intrigued by 3D printing, but few know how best to use it.

As with any new technology, 3D printing needed "early adopters" to take the lead on implementation. The medical field, for example, became one of them, taking notice of 3D printing in the 1990s. By 1999, a group at the Wake Forest Initiative for Regenerative Medicine had printed the first

Figure 4.1
Though the first 3D printed object in space was a simple faceplate engraved with names, it was also an important test of new technology in microgravity that would help prove the viability of an on-demand "machine shop" in space. Astronauts could then load up 3D files to make a special screwdriver or perhaps a replacement part for their coffee machine to solve immediate problems that can arise when far from home or the nearest hardware store. This 3D printer was photographed in 2014 after successful testing and certification, prior to its launch to the International Space Station.

artificial organ: a bladder created from a CT scan and layered with cells from the patient. In the following years, kidneys, livers, and other organs have been generated using a combination of human cells and 3D printing.

By the turn of the century (and millennium), many people were eager to make 3D printing more accessible. The "maker movement," those who embrace a do-it-yourself and inclusive approach to building or making, played a big role in this. While some in the maker community focused on hardware, others worked on making free or inexpensive software to interface with 3D printers more easily. In a relatively short time, 3D printers moved from large, expensive commercial machines to something that could be purchased in a local box store or even assembled from parts off the shelf and controlled from a personal computer.

Some in astronomy paid attention to the evolving landscape of 3D printing. What are the potentials for scientists to learn about their data in different ways through 3D printed models? And, perhaps even more important for a predominantly visual field, how can we share what we have found with new audiences? For example, could we reach the more than 250 million estimated people around the world from blind and low-vision communities?

While there have been projects that create tactile versions of astronomical data for people who are blind or visually impaired, 3D printing allows for another step. Early studies have shown that using 3D prints of astronomical objects promotes learning gains in astronomy and general spatial-skill recognition among other positive outcomes.

In short, 3D printing may be able to reach people that may have otherwise largely been shut out of astronomy, and it can benefit many groups who may be interested in learning about space. The ability to hold the 3D model of a dead star in your hands, for example, is arguably a much different experience than reading about it in a book or looking at data on a computer screen. Research suggests that using different senses such as touch to learn about space science topics does increase interest in such fields and increases the learner's enjoyment, particularly for underrepresented groups in STEM education.

As discussed, the 3D Universe can reach the physical world through printing, but it also extends the reach of these data further into digital space as well. 3D models can take on other levels of interactivity in a virtual world, specifically through virtual reality. Defined as a technology that simulates a person's physical self in a computer-generated environment, virtual reality is a concept (and buzzword) that has been around since the mid-twentieth century.

Despite its catch-phrase worthiness, virtual reality (VR) as a product didn't become much of a consumer option until the 1990s. Soon after that, however, VR got a big boost when some industries—including gaming—realized its potential for their commercial markets. Scientists, including astronomers, once again took notice.

Some early applications of astronomy in VR include simulated worlds based on data of Solar System objects, and later exoplanets, which are planets orbiting stars other than our Sun. These latter experiences were artist's impressions of what the surface of some of these distant and foreign worlds might be like. (Almost all the early exoplanets discovered in the late 1990s and through the next decade were "hot Jupiters," massive planets in extremely close orbit to their host star, making them very hot. As of the present day, astronomers have found thousands of exoplanets that span a range of physical properties—including those that may closely resemble Earth in some key ways.)

Meanwhile, planetary scientists have sent robotic probes to destinations around the Solar System just within the past decade, exploring everywhere from the hellscape of the surface of Venus to the rocky terrain of Mars. We have also had a probe drop through the atmosphere of Titan (one of Saturn's moons), fly around Jupiter, visit the far side of our Moon, and plunge closer to the Sun than anything built by humans before. And there are far more missions in the pipeline to explore other corners of our Solar System. The spectacular data these missions return to Earth can be used in virtual reality to allow anyone—scientist or not—to roam through our Solar System as never before.

The technology of extended human reality also keeps developing. In recent years, "augmented reality" has become popular as it layers more information (text, sound, images) onto real or virtual environments. These "AR" experiences are becoming easier to access as more smartphones, wearable computer devices, and other hardware support it. Once again, the technology has opened the door for exploring the Universe and the data we've gathered about it in new ways. Today, you can drive on the Martian surface with the Curiosity rover from your living room floor, launch the latest SpaceX spacecraft on a dining table, and walk through the debris of an exploded star—all in AR.

VIRTUAL AND AUGMENTED REALITY AND HOLOGRAMS

As far as the 3D Universe has come, there is still much further to go. Holograms are another concept that has criss-crossed the line between science fiction and science fact. While holograms do not yet match the capabilities of those in *Star Wars* or *Star Trek*, they are already being used in many ways, including in astronomy.

What's the difference between VR, AR, and holograms? Perhaps most important, holograms let people explore data in 3D, but do not require headsets or hand gear like VR and AR typically do. The technology of holograms is also different, with holograms using laser beams to bounce and reflect light of captured 3D objects and sensors. This allows a device to detect the viewer's motion so the user can interact with the 3D object. There are some astronomical datasets, such as certain supernova remnants, which lend themselves very well to the capabilities of holograms. More examples of astronomical objects being presented as holograms are in the pipeline, so look for them in the coming months and years.

It's not hard to get caught up in the technology and the gee-whiz factor of exploring space in these virtual environments. But there are also tangible educational benefits if the products take advantage of their platforms to deliver information that otherwise can be tricky to absorb. Concepts like scale of distance or size or how light behaves can all be communicated in innovative ways in a computer-generated 3D world that is much different from the 2D equivalent on a page or screen. And scientists are learning new ways of experiencing their own data, which may enable new ways, or perhaps more efficient ways, of making discoveries.

Whether it is transforming a 3D dataset into a physical form of printing or taking it further in the digital realm through extended reality, the 3D Universe is expanding (pardon the pun with what's happened in the Universe since the Big Bang). Now it's time to explore examples of the 3D Universe.

5

GETTING ORIENTED

Before we begin our 3D exploration of space and the marvelous machines that allow us to do so, let's start closer to home.

With the issues and distractions of the modern world, it is very easy to lose our inherent and ancestral relationship with space. Yet our interest in the Universe dates back millennia and spans the globe. The Aboriginals and Torres Strait Islanders developed astronomical practices to solve problems involving calendars, navigation, and weather; almost three thousand years ago, Chinese astronomers made accurate records of a solar eclipse and discovered a comet; Native American tribes in North America used the stars and their motions to help plan their activities and develop a system of beliefs.

Constellations from any culture can help us make (or remake) the connection that we as humans feel to space. While they may be on some levels simply star patterns connected to stories and folklore, they also can serve as a path to feeling more familiar and comfortable with our place in the Universe.

Of course, contemporary society poses new obstacles to our connection with the night sky that our ancestors could hardly imagine. Light pollution from artificial lighting in our industrialized world diminishes the view of the night sky. According to a 2016 study, more than 80 percent of people on the planet live under "skyglow," where the sky is just inherently brighter due to collective light humans send into the sky. A relatively new impediment to fully seeing stars, planets, and beyond from the surface of the Earth is the advent of large arrays of satellites that have been launched for communications and other purposes. In addition to interrupting the view of professional astronomers, these satellites can streak across the sky and disrupt our unfettered view of it.

Despite these challenges, there are still many ways for us to reclaim our cosmic connection. The chapters that follow outline some of the incredible achievements humanity has made in the quest to understand our Universe. And the night sky is still there, even if its wonders have been dimmed in many locations.

To connect us back with our shared night sky, we have selected a small number of 3D printable constellations for this book that were created by researcher Bruce Bream as part of the Star Coin Project. Some of these constellations take up large portions of the sky and include some of the brightest stars that can pierce the veil of artificial light. There are many other constellations, including the eighty-eight constellations recognized by the International Astronomical Union (IAU), the worldwide governing body responsible for astronomy.

The IAU-sanctioned constellations are attributed to ancient Greece, but they are thought to have built on the work done by even older Babylonian, Egyptian, and Assyrian cultures. There are also completely different collections of constellations with stories from peoples on continents around the globe. Each is to be celebrated and validated. And perhaps the most important stories about the night sky are the ones that we make up on our own after time spent gazing at the heavens.

Feel free to start your 3D voyage with this handful of "official" constellations and jump off into your own journey of discovery. As the following pages will show, there is so much out there to uncover.

CASSIOPEIA

In Greek mythology, this "W"-shaped group of stars represented queen Cassiopeia of Ethiopia. The constellation was also identified by the Egyptians, but with an association to an evil god. The Chinese saw a rider on a chariot, while these stars formed the king of the Fairies to the Celts.

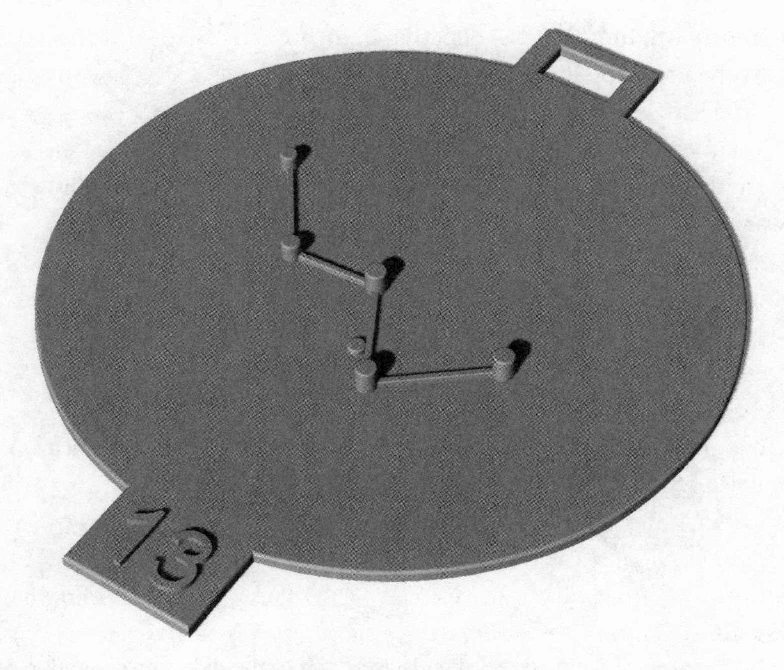

Figure 5.1
Cassiopeia's constellation is zigzag shaped like a "W" or an "M," depending on its orientation in the sky (or a viewer's orientation on the ground). This Northern Hemisphere constellation is in a circumpolar position, so that it seems to revolve as if it is centered on Polaris, the pole star.

The main shape of the Cassiopeia constellation is made up of five stars that are bright due to their relative proximity to Earth, all within a few hundred lights to us. This group of stars is visible all year long in the Northern Hemisphere and in the northern part of the Southern Hemisphere during spring. The Cassiopeia A supernova remnant detailed earlier in this book is found within the constellation, but you cannot see it without a powerful telescope. The extra star on the right side of the W is Eta Cassiopeia (η Cas), a binary star system.

URSA MAJOR

The Greeks and some Native American tribes saw this collection of stars as representing a walking bear, though with differing details associated

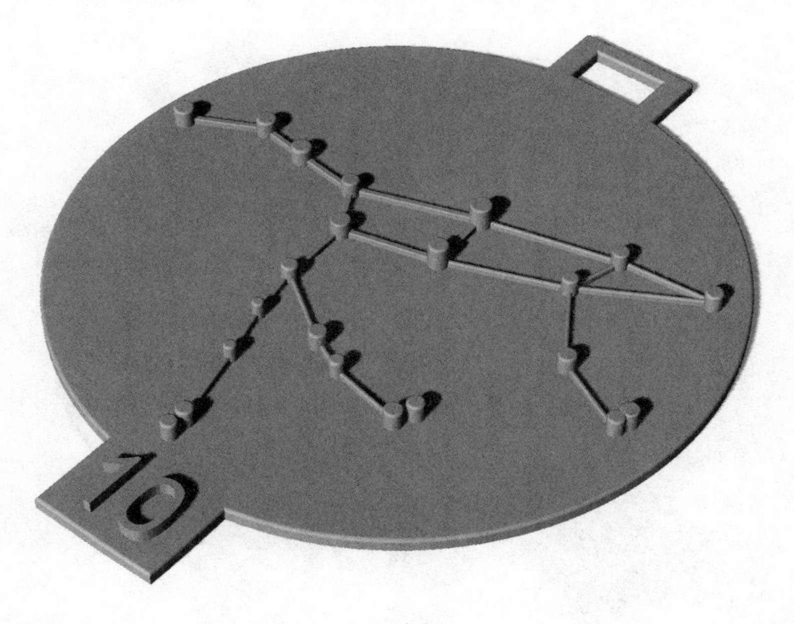

Figure 5.2
This Northern Hemisphere constellation is likely one of the oldest recognized in our night sky. A common pattern seen in Ursa Major is an asterism, or a smaller grouping of the brightest stars in the larger constellation, which outlines what many refer to as the Big Dipper.

with the stories. Some think that Ursa Major is one of the oldest constellations in the sky, with evidence of its existence dating back to prehistory.

Ursa Major is perhaps most famous for the asterism (that is, a prominent pattern or group of stars that is not actually a constellation) of seven stars encompassing the Big Dipper. It is one of the most spread-out constellations in the sky, covering about 1,200 square degrees. By contrast the full Moon only covers 0.2 square degrees. Ursa Major is visible most of the year from the Northern Hemisphere.

SAGITTARIUS

This constellation likely descended from Sumerian, Babylonian, and perhaps Arabian origins., The Greeks associated this pattern of stars with an archer. Sagittarius is part of the zodiac, the twelve constellations that the

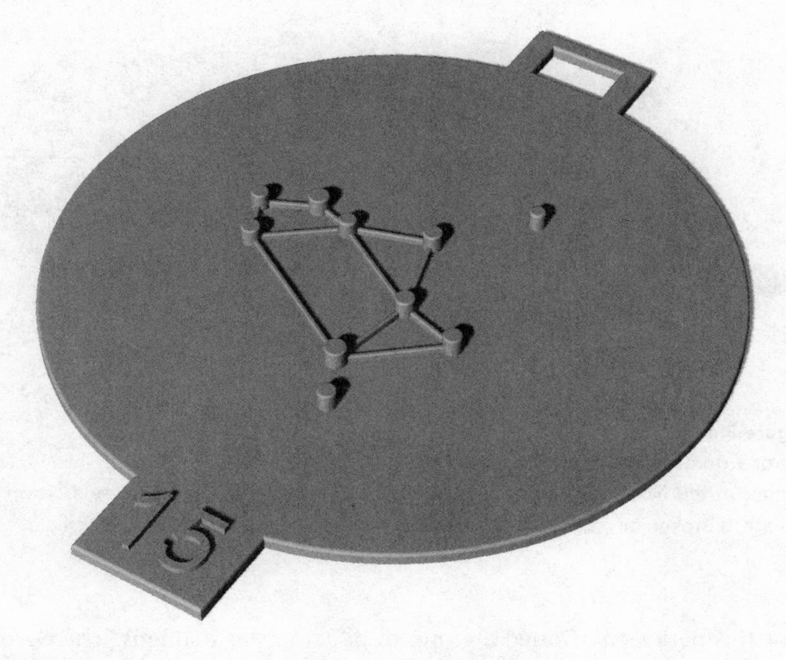

Figure 5.3
Sagittarius is one of the Zodiac constellations and visible in both hemispheres. These twelve constellations may be some of the oldest patterns in the night sky recognized by human civilizations.

Sun, Moon, and planets appear to move through. There is also a well-known teapot-shaped asterism within Sagittarius.

The constellation of Sagittarius contains seventeen named stars, including seven that are rather bright and therefore relatively easy to spot with the unaided eye. There are many distant objects near or around this constellation, but perhaps none more intriguing than the center of the Milky Way where our Galaxy's supermassive black hole resides. The constellation even lends this important object its name, Sagittarius A* (Sgr A*).

CORONA BOREALIS

Given its shape, it may not be surprising that Aboriginals saw this pattern of stars as a boomerang. Other civilizations like the Shawnee tribe of

Figure 5.4
Corona Borealis is a relatively simple open circlet of stars recognized by many civilizations in the Northern Hemisphere. It has been identified or pictured as a crown or wreath, a broken or cracked dish, a boomerang, and a group of dancing star maidens.

North America envisioned a group of dancing star maidens. The Greeks thought Corona Borealis, as they called it, was a crown or a wreath.

The name means "northern crown" in Latin so it should not be surprising that this constellation is found most easily from the Northern Hemisphere. The brightest star in Corona Borealis is Alphecca, which is a double star system. There are other intriguing objects associated with Corona Borealis ranging from individual stars to galaxy clusters, but a large telescope is required to spot them.

CENTAURUS

In Greek mythology, Centaurus is a centaur, a hybrid of a man and horse. Centaurus's collection of stars is quite spread out across the sky, making it one of the largest constellations. It also contains Alpha Centauri and Beta Centauri, two of the brightest stars in our night sky. Within this constellation, you will find Proxima Centauri, the closest star to the Sun at just a shade over four light-years away. And within this patch of the sky, there is also a famous globular cluster, Omega Centauri or NGC 5139, and a well-studied galaxy known as Centaurus A, or NGC 5128, which has a powerful jet shooting out its center.

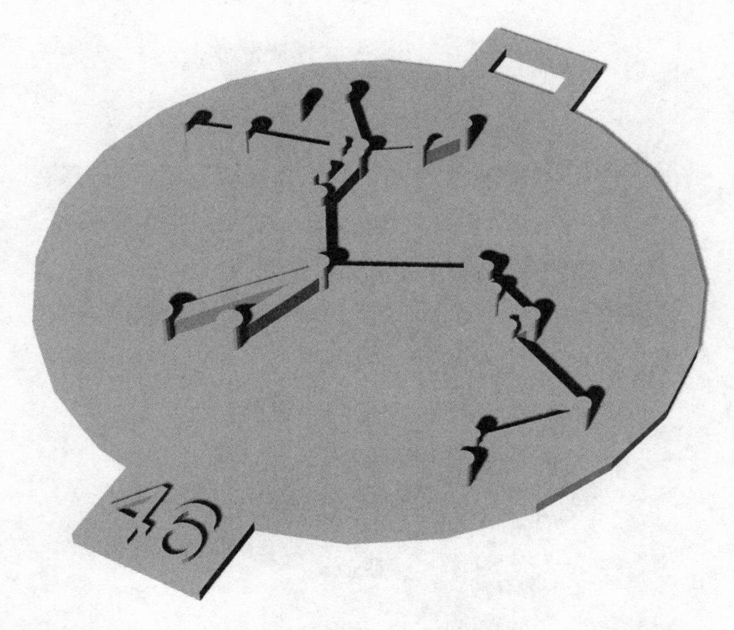

Figure 5.5
A complex Southern Hemisphere constellation, Centaurus is one of two constellations recognized in the night sky to represent a centaur, a mythical creature consisting of a human head and torso on a horse's body. The constellation has a large, four-sided shape imagined by the ancient Greeks to be the human upper part, attached to its two horse legs.

The constellation of Centaurus is mainly visible from the Southern Hemisphere. Because of its location in the sky, it does not have "M" or Messier objects because it was too far south for Europe-based Charles Messier to observe it. During his lifetime in the eighteenth and early nineteenth centuries, Messier catalogued hundreds of astronomical objects that still bear his name.

PEGASUS

The Greek astronomer Ptolemy described the constellation of Pegasus nearly two thousand years ago. Named after the mythological winged horse, the constellation of Pegasus has fifteen formally named stars (that is, those recognized by the International Astronomical Union). While there is one

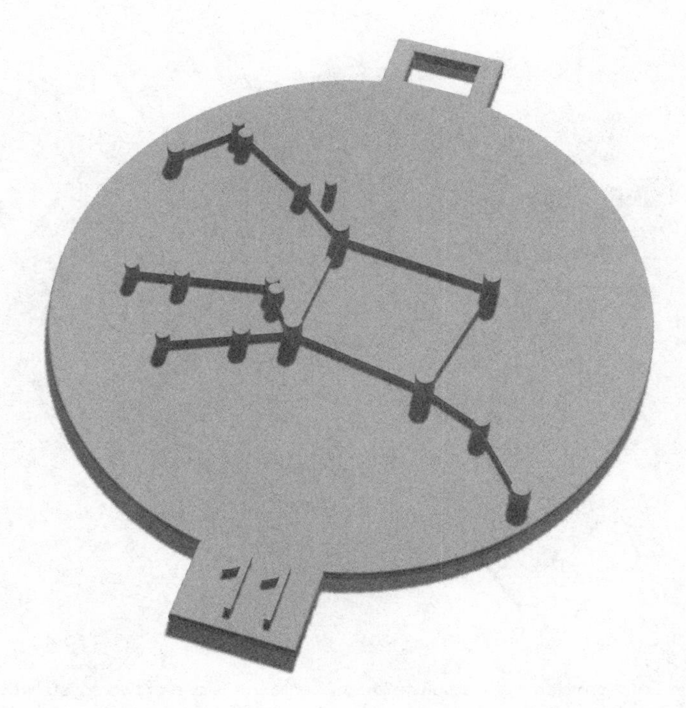

Figure 5.6
Visible in both hemispheres, Pegasus has fifteen named stars and in Greek culture was thought to resemble the upper part of a winged horse. It has a strong square or box shape on one side and is also identified in Chinese astronomy as part of the Black Turtle constellation, in Hindu as a lunar mansion, and a moose for the Ojibwe tribe of North America.

Messier object within Pegasus, the globular cluster M15, there are many more sources from the NGC catalogue. This abbreviation stands for "New General Catalogue," a compendium of stars, star clusters, nebulas, and galaxies assembled by the Danish and British astronomer John Louis Emil Dreyer and published in 1895.

There are certainly many interesting cosmic objects tied to this constellation, but perhaps none more so than 51 Pegasi. This is a star located about 50 light-years from the Sun that is like our home star in size and age. Even more important, astronomers discovered a planet orbiting this star in 1995. This was one of the earliest planets discovered outside of our Solar System, which are now commonly referred to as "exoplanets."

6

THE EXPLORERS IN 3D: TELESCOPES, PROBES, AND ROVERS

The machines that humans have produced to explore space are remarkable. While some build on existing and previously proven technology, each new facility is designed to push the boundaries of science. This generally means that new pieces of hardware need to be imagined, developed, and constructed. And in the case of many, they need to be lofted into the harshness of space and continue to work without any physical upgrades or repairs from anyone here down on Earth.

Since the dawn of the Space Age, hundreds of telescopes, satellites, and other instruments have been launched to help us explore our Universe. Some of them barely skirt Earth's gravity and are expected to operate for a short time. Others travel farther and can sometimes fruitfully work in space for decades, such as the Hubble Space Telescope that went into orbit in 1990 or the Voyager spacecraft that have been traveling through the Solar System since the mid-1970s.

Regardless of their duration of scientific usefulness, each of these robotic space explorers is an engineering marvel. There are no giant factories that churn these out. Most of these projects were green-lighted and funded because they had the potential to uncover new information in a way that hadn't been done before. Teams of engineers, scientists, technologists, administrators, and others must collaborate across government, academia, and industry to take each new project from ideas and hopes to machines and software.

The 3D engineering models of these telescopes, probes, and rovers are amazingly complex. The versions of the 3D models covered in this book, however, are simplified. This allows us to gain a sense of how each machine is shaped and highlight some of its key features without being overwhelmed with an abundance of technical details.

In this chapter, we will explore these spacecraft and telescopes in 3D in the chronological order of their operational debuts. Later in the book, we will find out how some of these marvelous machines have, in fact, returned data that allow us to build 3D models of cosmic objects.

Authors' Note

The 3D models in this chapter and throughout the rest of the book were typically the result of the work of many people. For example, there are countless scientists, engineers, and others who conceived of and developed the 3D models that would eventually become the physical manifestations of these spacecraft. Similarly, the 3D models of astronomical objects in the chapters that follow often represent the efforts of a team of individuals who have contributed their intellect, talent, and energy (as well as observing time from telescopes) to create them. The credits in this book list only a representation of the individuals involved in the pipeline of conceptualizing and developing these models over time. We invite you to visit this book's companion website at https://starsinyourhand.pubpub.org/ for more information on each 3D model.

SATURN V

To date, the Saturn V rocket (the "V" is the Roman numeral for five) is the only rocket that has delivered humans to the Moon. As tall as a thirty-six-story building, the Saturn V was used in the Apollo missions of the 1960s and early 1970s as well as for launching the Skylab space station in 1973. In total, it flew thirteen flights including ten with people on board. Developed at the height of the Space Race with the Soviet Union, the Saturn V went from design to flight in just six years. NASA is currently working on its next generation of heavy lift rockets, known as the Space Launch System, which bears many resemblances to the Saturn V.

The Saturn V rockets in the Apollo missions had three stages (the one that launched Skylab only had two). The first stage was the most powerful as it did most of the work of overcoming Earth's gravity and getting the rocket off the ground. It also burned through a remarkable amount of fuel—about 40,000 pounds (about 18,000 kilograms) per second in its first two minutes after ignition. The second stage got the Saturn V almost into orbit, while the third stage put the Apollo spacecraft into Earth's orbit and on its way to the Moon.

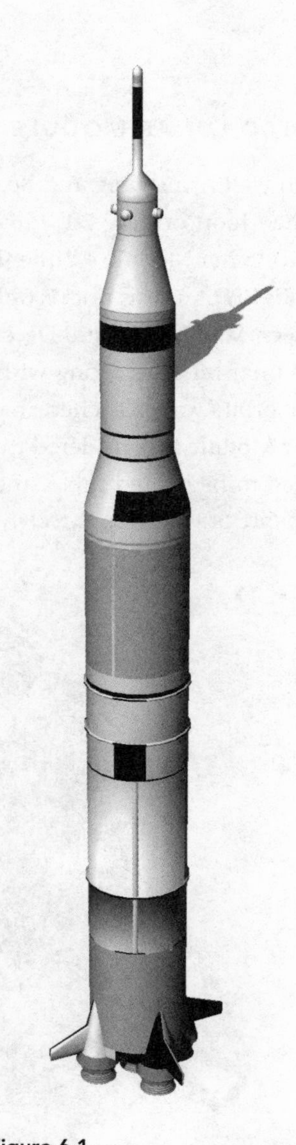

Figure 6.1
The Saturn V rocket flew thirteen flights and successfully delivered humans to the Moon. Standing about thirty-six stories tall, it had three stages to lift the rocket off the ground, boost it to near orbit, and then deliver it into Earth's orbit and beyond to the Moon.

APOLLO LUNAR MODULE

When Neil Armstrong and Buzz Aldrin became the first humans to land on the Moon on July 20, 1969, the vehicle that got them there was the Apollo Lunar Module. While this was certainly the Lunar Module's most famous flight, it was not its only one. The Lunar Module made ten flights between March 1969 and December 1972 with six landings on the Moon. The Lunar Module, along with the Command Module that remained in lunar orbit, was launched into space aboard the Saturn V rocket. The Lunar Module is considered the first spacecraft to operate in space, but it needed to be carried there since it was structurally and aerodynamically incapable of traveling through the Earth's atmosphere.

Figure 6.2
The lunar vehicle that brought the first humans to the Moon had a bug-like shape, four legs and large feet, a mylar body, and two parts of the main module for descent and ascent.

The Lunar Module had an unusual, almost bug-like shape. Its four legs were spread apart and equipped with big "feet" so it wouldn't sink on what was then an unknown lunar surface. The main body of the Lunar Module was covered in mylar—like the material of a popular type of balloon and emergency blanket—to keep the heat in and protect it against tiny rocks or other debris. The module is made up of two main pieces: the descent-stage part on the bottom and the ascent part on the top. Each of these was equipped with an engine, the former to lower the astronauts to the lunar surface and the latter to return them to the command module.

VOYAGER

In 1977, NASA launched twin spacecraft, Voyager 1 and Voyager 2, from Cape Canaveral, Florida. The primary mission for the Voyager spacecraft was to study Jupiter and Saturn, which they did, but there was so much more to do. In their over four decades in space, the Voyagers have

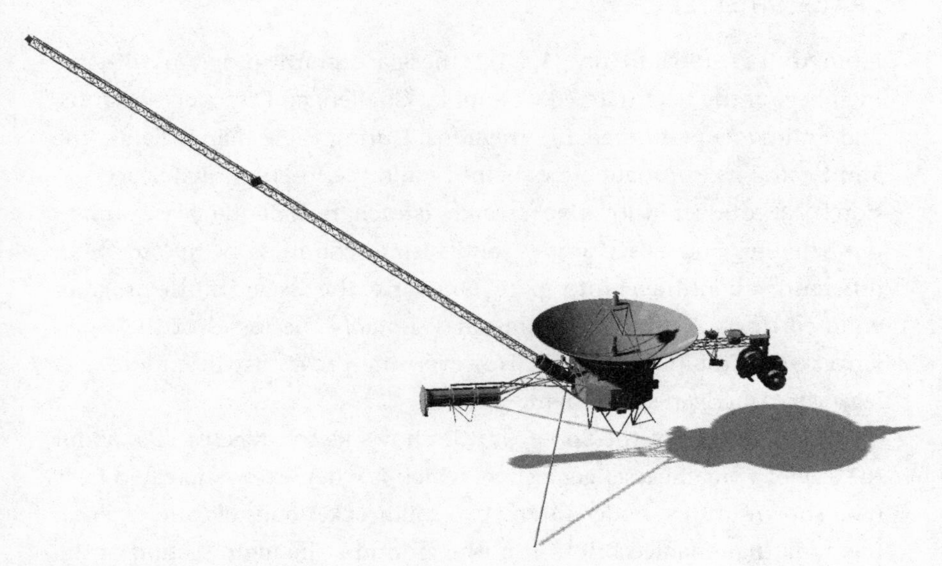

Figure 6.3
The two Voyager spacecraft are identical to one another. The long boom that extends from Voyager is a magnetometer, measuring magnetic fields of objects in the Solar System. The central large disk is the high-gain antenna that enables the spacecraft to transmit data to and from Earth.

explored all the giant planets of the Solar System including not only Jupiter and Saturn, but also Uranus and Neptune along with forty-eight of the moons of these planets. Today, the Voyagers are still collecting data as they head farther than any human-made object has, now exploring the region beyond the known planets and to the outer limits of the Sun's influence. Both Voyagers carry a copy of the "Golden Record," which contains images, sounds, music, and samples of fifty-five languages from Earth in the hope that an intelligent species will encounter them.

The two Voyager spacecraft are identical. The long boom that extends from Voyager is a science instrument known as a magnetometer, which measures magnetic fields from the Sun and other bodies in the Solar System. The central large disk is the high-gain antenna that allows each Voyager to transmit data back to Earth and receive commands from home. There is a total of ten science instruments aboard the Voyagers that are sensitive to different kinds of light including radio waves, ultraviolet, and infrared.

SPACE SHUTTLE

From April 12, 1981, to July 21, 2011, the Space Shuttle program, with five members of the fleet (named Columbia, Challenger, Discovery, Atlantis, and Endeavor) performed 135 missions. During these many flights, the Shuttle and its astronaut crews helped build the International Space Station (ISS), conduct innovative research, launch, repair and release numerous satellites and observatories, and inspire countless people to think differently about travel into space. Of course, the Space Shuttle program endured many obstacles and tragedies, notably the losses of the Challenger and Columbia shuttles. However, the program will endure as a legacy to American space flight.

This 3D model of the Space Shuttle shows its so-called orbiter, about the size of a small passenger airline, which is where crew and cargo flew. (Not shown in this model are the two solid rocket boosters and external fuel tank that enabled liftoff from the ground.) The main section of the orbiter contained the cargo bay, which was large enough for a robotic arm to help deploy and grab sections of the ISS or large observatories

Figure 6.4
NASA's Space Shuttle program included 135 missions and five vehicles during its lifetime from 1981 to 2011. This 3D model shows the Space Shuttle's orbiter, which is about the size of a small passenger airline, where the crew and cargo flew. The cargo bay, shown with its doors opened, would typically house supplies or telescopes to be launched.

including the Hubble Space Telescope. While it launched like a rocket, the Shuttle was designed to land like a glider, which is why it was given small wings and its airplane-like shape.

HUBBLE SPACE TELESCOPE

Arguably the most famous telescope ever built, the Hubble Space Telescope has captured the hearts and minds of the public and scientists alike since its launch in April 1990. Although the telescope suffered a major setback when scientists and engineers discovered that the primary mirror had a flaw (technically known as a "spherical aberration," meaning it could not properly focus light), a dramatic mission by astronauts in December 1993 fixed the problem and restored Hubble's sharp vision. Hubble was the first astronomical observatory to be placed into orbit around the Earth and completes one orbit every ninety-seven minutes.

Figure 6.5

An astronomical icon, Hubble has a primary mirror that is almost 8 feet (about 2.4 meters) in diameter, with the spacecraft measuring about 43 feet (13.2 meters) long. The observatory uses four reaction wheels to orient itself. To keep the mission powered, Hubble's solar panels capture and transfer energy to be stored on board.

Hubble's main, or primary, mirror is almost 8 feet (2.4 meters) in diameter. While much larger telescopes on the ground have been constructed, Hubble's mirror is unaffected by the distorting effects of the Earth's atmosphere. This allows Hubble to see both exquisite details and extremely faint features in objects across space. The spacecraft is about 14 feet (4.2 meters) long with four reaction wheels in the middle used to orient the observatory. The solar panels transfer energy to batteries on board. NASA has conducted five separate missions with astronauts to repair and upgrade Hubble, but no more are planned with the end of the Space Shuttle program in 2011.

CASSINI-HUYGENS

On October 15, 1997, a rocket blasted into space carrying not one but two very special pieces of cargo. Stowed inside the Cassini orbiter, on a NASA mission, was the European Huygens probe. Together, the pair would spend the next seven years traveling to the Saturn system. Once there, Cassini orbited the famous ringed planet for thirteen years, sending data back to Earth. On January 14, 2005, Huygens was released from inside Cassini and the probe descended to Titan, one of Saturn's moons. Thus, Huygens became the first human-made object to land on a world in the outer Solar System. It sent back invaluable data of Titan's atmosphere and its surface before it stopped sending signals back to Earth after it impacted Titan's surface as expected. Together, Cassini and Huygens allowed us to observe seasons and weather on another planet, explore an extraterrestrial ocean, and expand the possibilities of what exists in the outer worlds of our Solar System.

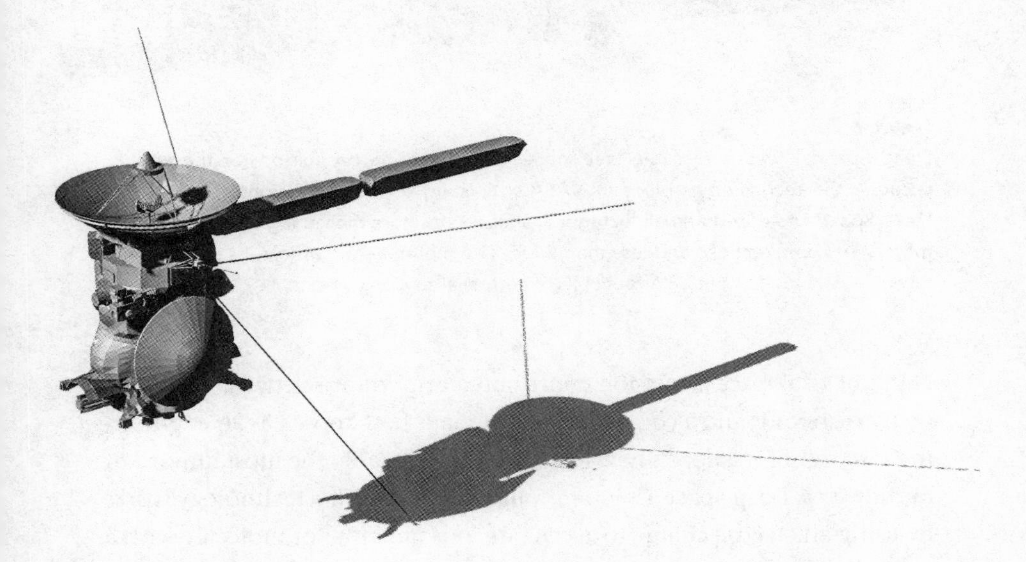

Figure 6.6
The Cassini-Huygens mission was to explore Saturn and one of its moons, Titan, via an orbiting observatory (Cassini) and a space probe (Huygens). The spacecraft duo had numerous instruments on board built to retrieve information on the atmospheric, oceanic, and other conditions of the ringed planet and its largest moon.

Cassini-Huygens was a veritable full science laboratory launched into space that traveled billions of miles to the Saturn system. Between the Cassini orbiter and the Huygens probe on board, they carried instruments that collected information in different kinds of light (visible light, infrared, ultraviolet, and radio waves), and others that could detect magnetic fields, dust, plasma, and more. Most of these were inside the main body of the spacecraft and wrapped in thermal blankets to protect them from the hazards of space. The large radar disk on the outside allowed Cassini to explore Saturn and some of its moons.

DEEP SPACE 1

Figure 6.7
Deep Space 1 was one of the first spacecraft to use ion propulsion for the engine, securing the technology's place in NASA's planning for other missions. The body of Deep Space 1 featured a small, octagonal-shaped structure measuring only 3.6 feet (1.1 meters) in two directions and less than 5 feet (1.5 meters) at its widest. Its solar panels, however, spanned about 38.5 feet (11.75 meters) after deployment.

NASA's Deep Space 1 mission contributed important science after it flew by two asteroids and a comet, but it is perhaps best known as an engineering test flight for a dozen new technologies. Arguably the most important of those was Deep Space 1's ion propulsion engine. This technology works by using an electric charge to accelerate ions (an atom or molecule with a net negative charge) from xenon fuel. This process releases electrons from the xenon and creates positively charged ions, which are accelerated out of the thruster as a beam that moves the spacecraft. The success of the ion propulsion engine in Deep Space 1 led to NASA using the technology in other missions already launched and more planned for the future.

The main body of Deep Space 1 was an octagonal-shaped aluminum structure. Deep Space 1 can be considered a "mini-satellite" since it measures just 3.6 feet (1.1 meters) in two directions and just under 5 feet (1.5 meters) at its longest. The solar panels spanned about 38.5 feet (11.75 meters) when fully deployed. The ion propulsion system vented out of the small circular opening on one end of the octagon, which moved Deep Space 1 to its destinations once it escaped Earth's gravity via a more powerful rocket. The spacecraft's antenna for communicating with Earth was located on the opposite end.

CHANDRA X-RAY OBSERVATORY

Figure 6.8
The Chandra X-ray Observatory has three major parts: the X-ray telescope containing mirrors that capture X-rays from objects in space, a collection of science instruments that record the X-rays, and the spacecraft that houses the telescope and instruments. About the size of a school bus, Chandra was the largest and heaviest payload ever to be fitted in the Space Shuttle for launch.

While the Hubble Space Telescope is certainly the most famous, it is just one member of an extraordinary telescopic family. In the 1980s, NASA commissioned the "Great Observatories," each of which would detect different types of light. The Chandra X-ray Observatory was launched less than a decade after Hubble and has served its role as a premier explorer of X-rays in the Universe. Unlike Hubble, Chandra flies in a highly elliptical orbit that takes it much farther away from Earth and does not allow for servicing from astronauts. Therefore, for more than twenty years, Chandra has been performing groundbreaking research without upgrades or repairs, making discoveries about areas from black holes to exploded stars to dark matter.

One of the most innovative features of school-bus-sized Chandra is its mirrors, which are unseen in this 3D model but extend nearly the full length of the body of the spacecraft. Because X-rays are so energetic, they cannot be focused by traditionally shaped mirrors that simply absorb them. Therefore, Chandra's mirrors are a series of nested barrels, each carefully polished to have the incoming X-rays skip off them like a stone on a pond. The X-rays are then focused onto science instruments at the end of the telescope, some 30 feet (10 meters) away. Other components of the Chandra spacecraft include the sunshade where the X-rays enter and the solar arrays that provide electricity for its X-ray studies of the Universe.

DEEP IMPACT

On July 4, 2005, the Deep Impact spacecraft slammed into comet Tempel 1 less than six months after it was launched from Cape Canaveral in Florida. Not only did Deep Impact travel 267 million miles (429 million kilometers) to this icy relic from the primordial Solar System, but it also released an "impactor" to expose material on the comet's surface. The impactor was a separate, small battery-powered spacecraft. Weighing 815 pounds (370 kilograms), it flew on its own in a one-way final mission to obliterate itself against the comet. The material kicked up from this collision with the comet provided evidence of water ice and organic materials. This buoyed the idea that comets once delivered these essential compounds to Earth in the Solar System's early days.

For a relatively small spacecraft, Deep Impact packed quite a big punch (literally). At about 10.5 feet (3.2 meters) wide, the impactor spacecraft is about the size of a minivan. The largest flat piece is the solar panel, and the large dish is the antenna that was able to transmit data back to Earth in near real time. The two cylindrical structures in the 3D model are cameras with different resolutions. The Deep Impact mission also carried debris shields that would allow the spacecraft to

Figure 6.9
After the spacecraft slammed an impactor into a comet, the Deep Impact mission helped scientists solidify the idea that comets were essential in the development of life on Earth. About the size of a minivan, the spacecraft had a large flat solar panel, a large antenna to send data to Earth, and cameras (the two cylindrical structures), among other components.

pass through the inner coma—a cloud of gas and particles that surrounds the solid core—of the comet without being damaged enough to thwart the mission.

KEPLER

Before the mid-1990s, the idea of planets around stars other than our Sun was science fiction and speculation. By the time we reached the end of the millennium, scientists using relatively small telescopes on the ground discovered the first suite of "exoplanets" through the tiny wobbles that the planets caused in their host star. But scientists wanted other ways to find exoplanets and soon turned to searching dips of light coming from stars when planets pass in front of them. The Kepler mission looked for these "transits" from thousands of stars in space from its vantage point in space outside the blurring effects of Earth's atmosphere. From the start of the mission in 2009 until the spacecraft ran out of fuel in 2018, Kepler surveyed our Milky Way galaxy and found thousands of possible exoplanets, including some about Earth's size or smaller. Even though Kepler

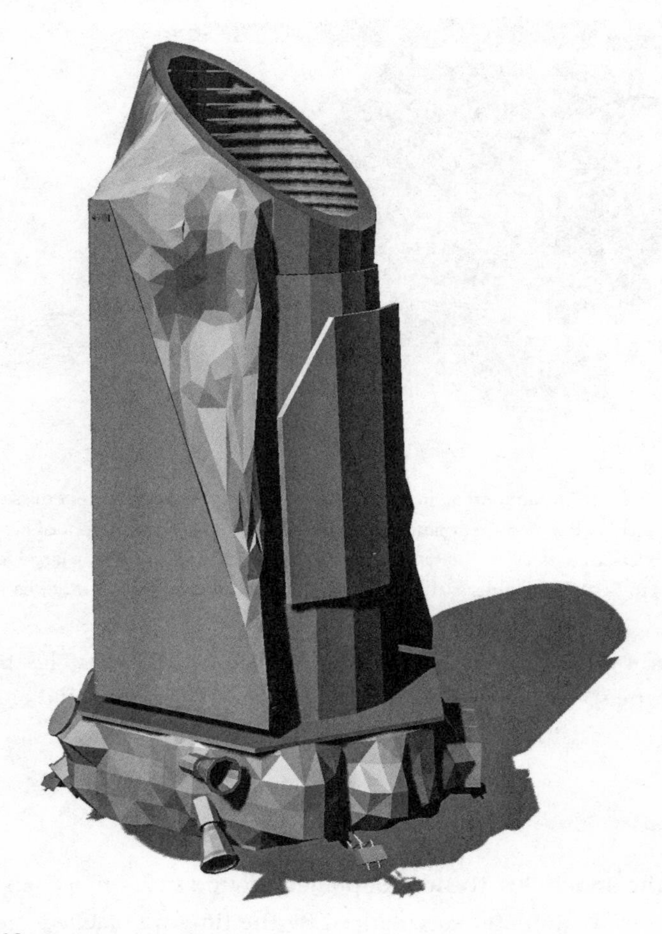

Figure 6.10
For nine years, Kepler helped scientists discover thousands of exoplanets, advancing research on new planetary bodies well beyond our own solar system. Kepler included a 55-inch (1.4-meter) telescope, a camera, the spacecraft that housed the components and provided power, a sunshade to block much of the Sun's light, and star trackers (two little horn-shaped objects on the opposite end from the sunshade) to orient the spacecraft.

is now retired, its data archive is a treasure chest in which we can keep digging for more information about new worlds.

The heart of the Kepler mission is a 55-inch (1.4-meter) visible light telescope designed to capture a relatively wide field of view. The light is then sent to a camera with forty-two charge-coupled devices, or CCDs, which are like cameras available in most smartphones. This single-purpose instrument is housed inside the spacecraft that provides power and, when it was on its original mission, the ability to scan different parts of the sky. The angled end of the spacecraft is a sunshade that blocks some of the light from the Sun so Kepler can focus on more distant stars. The two little horn-shaped objects on the opposite end from the sun-shade are star trackers that help the spacecraft orient itself on the sky.

LUNAR RECONNAISSANCE ORBITER

The Lunar Reconnaissance Orbiter, or LRO, journeyed to the Moon in four days after it was launched in June 2009. In its first three years, LRO followed an orbit that took it close to both poles of Earth's only natural

Figure 6.11
The Lunar Reconnaissance Orbiter (LRO) is helping to map the surface of our Moon for potential crewed and uncrewed missions to Earth's only natural satellite for detailed study. LRO carries over 200 pounds (90.7 kilograms) of scientific instrumentation, from a camera to a laser altimeter to other hardware, as well as a large dish for communications with Earth, and three solar panels measuring 14 feet (4.27 meters) across.

satellite. After that, LRO moved to an elliptical orbit that allows it to pass closely to the lunar southern pole. There were two main goals for LRO: exploration for future human and robotic missions as well as data gathering for scientific discoveries. LRO mapped most of the surface of the Moon and found evidence for the presence of water in ice and rocks in the craters and plains on its surface.

The LRO carries over 200 pounds (90.7 kilograms) of scientific instrumentation, including a camera that takes images in visible and infrared light, a laser altimeter for measuring landing slopes and polar ice, and other hardware that can detect radiation in space that could be helpful for informing future human exploration of the Moon. In the 3D model, the large dish on the LRO is its high-gain antenna that allows communication with Earth. The three solar panels are 14 feet (4.27 meters) across and indicate the spacecraft's scale in the 3D model.

CURIOSITY

Since the 1970s, NASA has made exploration of Mars a top priority and has sent a series of robotic rovers to the Red Planet, in part to lay the groundwork for future human explorers. The Curiosity rover (previously known as the Mars Science Laboratory) left Earth aboard an Atlas V rocket on November 26, 2011. When it arrived at Mars, the Curiosity rover made it to the planet's surface by a complex landing sequence that included a giant parachute, a jet that slows its descent, and a bungee-like apparatus called a "sky crane." Since surviving that perilous drop, the car-sized rover has been exploring Gale Crater, a region where scientists think a meteor struck about 3.5 billion years ago.

With Curiosity, scientists and engineers wanted to build a rover that could travel greater distances than its Martian predecessors, Spirit and Opportunity. Each of Curiosity's six wheels has its own individual motor and can turn the rover in place a full 360 degrees if necessary. At the front of the rover, a robotic arm can reach out to the Martian soil with a suite of different instruments. At the opposite end of the rover from the arm is the nuclear power source for the rover, and the antenna-like structure on the top analyzes the chemical composition of vaporized material created by the robotic arm.

Figure 6.12
Curiosity, the car-sized rover currently exploring the Red Planet survived "seven minutes of terror"—for its remote human observers—as it dropped to the Martian surface via a parachute, rocket-powered jets, and a bungee-like apparatus. Curiosity has six heavy-duty wheels, a robotic arm packed with science instruments, a nuclear power source, and an antenna-like structure to analyze the composition of Martian material that the robotic arm vaporizes.

JAMES WEBB SPACE TELESCOPE

The James Webb Space Telescope (JWST) carries the hopes of a generation of astronomers. NASA has dubbed JWST, which will observe infrared light, as the "premier observatory of the next decade." While some have described JWST as the successor to Hubble, it is distinctive for many reasons. For example, JWST had to travel much farther than Hubble to do its work—too far for astronauts to reach it—at about 1 million miles (1.5 kilometers) from Earth. Astronomers expect JWST will allow them to study the glow from the Big Bang, look for planets around other stars, and teach us new things about our Solar System.

Many technological and engineering feats had to be achieved to make this mission possible. One of the most striking innovative features in the 3D model is the giant honeycomb mirror on its top. Over 21 feet (6.5

Figure 6.13
The James Webb Space Telescope (JWST) will be NASA's next flagship mission for gathering information on such topics as the history of the Universe to the formation of exoplanets to the evolution of galaxies. Its 21-feet- (6.5-meter-) wide honeycomb mirror is on the right side of this image under the supports that hold its primary and secondary mirrors. The sunshade roughly the size of a tennis court that will protect the telescope's temperature-sensitive observations from sunlight is visible in its folded position on the left.

meters) across, this mirror is made of eighteen separate segments that need to unfold, like unwrapping origami, once in space. The structure opposite the primary mirror holds the secondary mirror that reflects light into the instruments within the body of the spacecraft. The bottom of the spacecraft is dominated by the tennis-court-sized sunshade that has five layers and blocks and deflect heat from the Sun so JWST can carry out its temperature-sensitive observations.

USING 3D PRINTING TO BUILD IN SPACE

While 3D printing has opened many opportunities on Earth, NASA and other space agencies are looking into what it can do in space itself. NASA has plans to return humans to the Moon and ultimately transport them to Mars. How will these space travelers get the infrastructure they need to live and work on Mars, so far away from our planet? One proposal is to create large-scale 3D printers that could use the lunar or Martian soil with additives to create habitats, laboratories, and other structures that would be essential to life on another world. NASA is beginning partnerships with some private companies to explore how 3D printing may play a role in humanity's next steps into space.

NASA began experimenting with 3D printing when it installed the first device on the International Space Station (ISS) in 2014. The microwave-sized machine has been used to build optical fiber, a wrench, an antenna part, and more, with no significant adverse effects from being manufactured in microgravity. NASA hopes to launch a next-generation 3D printer in the coming years that will be paired with robotic arms. This new 3D printer will be designed to potentially repair existing satellites and build new structures in orbit.

7

THE SOLAR SYSTEM

Sometimes it is possible to take your most immediate surroundings for granted. This can even happen with our cosmic neighborhood, the Solar System. For those who like to look at the night sky, the Moon, Mars, Venus, and other Solar System bodies become some of our earliest connections.

Of course, we've only scratched the surface—literally—of a few bodies outside of Earth. Our Solar System is complex and diverse, consisting of the Sun, eight major planets, Pluto and the rest of its dwarf planet cousins, a raft of moons, as well as countless smaller objects like asteroids and comets. The more we learn about the Solar System the more wonders we find and fascinating questions we collect to try to answer.

The realm of 3D technology allows us to explore the Solar System through new eyes and—in the case of 3D printing—fingers. What does the surface of the Moon really look and feel like? What will future robots encounter when we venture to Europa to look for life? Does an asteroid have the same terrain as any planet elsewhere in our Solar System?

The legendary landing of astronauts on the Moon marked a beginning of physical human exploration beyond Earth. Yet, it remains our most distant reach physically as a species nearly five decades later. Since then, we have sent machines in our stead, some of which are described in chapter 5.

Our robotic explorers have scanned, analyzed, and driven over these other worlds and delivered data back to Earth that allow us to reconstruct what they have learned. The other bodies in the Solar System not only reveal themselves to be intriguing in their own right, but they also tell us about the history of our own planet Earth. Questions that arise include: How did the Solar System form? When and where did water exist? Could there be life—in whatever form—elsewhere?

It was not many decades ago that many people, lacking information discovered since, thought Mars could be home to existing intelligent life. Sending the Viking missions in the 1970s dispelled that branch of thinking as science fiction only. Each successive mission to Mars has brought us a bit closer to understanding the real story of the Red Planet and how it mirrors and differs from Earth's past.

The inner—or terrestrial—four planets are Mercury, Venus, Earth, and Mars. They are often categorized together because they all have hard surfaces containing rocks. Venus and Mars both once may have had more similarities to Earth than they do now, but too much atmosphere on the former and not enough on the latter sealed their very different fates. Meanwhile, gas giants of Jupiter, Saturn, Neptune, and Uranus look nothing like the planet we inhabit. They are primarily gas—hydrogen and helium as well as other heavier elements in nonsolid forms—with comparatively small rocky cores.

While the gas giants seem unlike anything we know here on Earth, some of their moons are more recognizable to us in certain ways. Saturn's moon Titan, for example, has oceans, rivers, and seas. (Unlike Earth, however, these bodies of liquid are made of methane.) Europa, one of Jupiter's largest moons, perhaps possesses an ocean of saltwater beneath its surface, making it one of the most compelling places in the Solar System to look for evidence of life. However, this possible saltwater ocean sits beneath a shell of ice that is more than ten miles (sixteen kilometers) deep—a challenge that any robotic explorer would need to overcome.

The smaller bodies of the Solar System are just as intriguing. There are about a million known separate bodies in the asteroid belt between Mars and Jupiter. Thought to be a remnant from the formation of the Solar System, the asteroid belt contains clues to the very origin of our cosmic neighborhood. Some also see monetary value in asteroids, and plans are underway to one day mine these bodies for the precious metals and minerals they are believed to contain.

Far beyond Pluto's orbit lies vast clouds of rocks in a range of sizes that constitute the Kuiper Belt and the ever more distant Oort Cloud. The Kuiper Belt is a doughnut-shaped region beyond Neptune that contains millions of objects, while the Oort Cloud is the most distant region of our Solar System and surrounds everything in it including the Kuiper Belt. Scientists think both the Kuiper Belt and the Oort Cloud are the source of comets, some of which come our way when occasionally nudged toward the Sun. Comets are planetary interlopers that may have delivered water to our planet during the infant days of our Solar System. These visitors represent the multitudes that exist at the Solar System's distant edges.

What can we explore in 3D in the Solar System? To date, the best places to travel virtually include three destinations: the Moon, Mars, and a handful of asteroids. (This makes sense when you think about both proximity and the many missions we have sent to these relatively nearby objects.) We have included several examples of places to visit in 3D on the Moon, Mars, and asteroids.

Beyond that, there is data about other Solar System objects that are available not as full 3D models, but as tactile topographic 3D prints that allow the user to physically feel the differences of the various surfaces. In some of these 3D models, the heights of the features have been exaggerated so the user can see and feel them more clearly. We note where this is done and the scale of the enhancement.

THE MOON

The lunar landscape is a mixture of bright highlands and dark "seas" once filled with lava, both of which now show the scars of large impact craters and rays of ejected material. Scientists think the Moon itself was formed after a violent collision with the Earth billions of years ago.

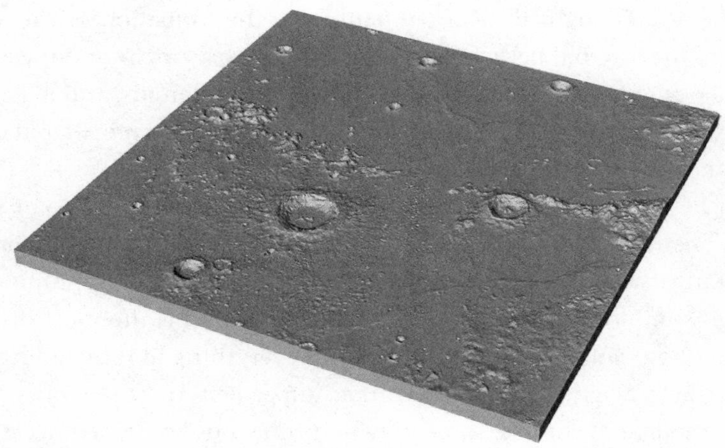

Figure 7.1
The near side of our Moon has a smooth and less pockmarked texture than the far side of the Moon. Instead, the near side is filled with large shallow basins due to giant lava flows that filled in many craters billions of years ago.

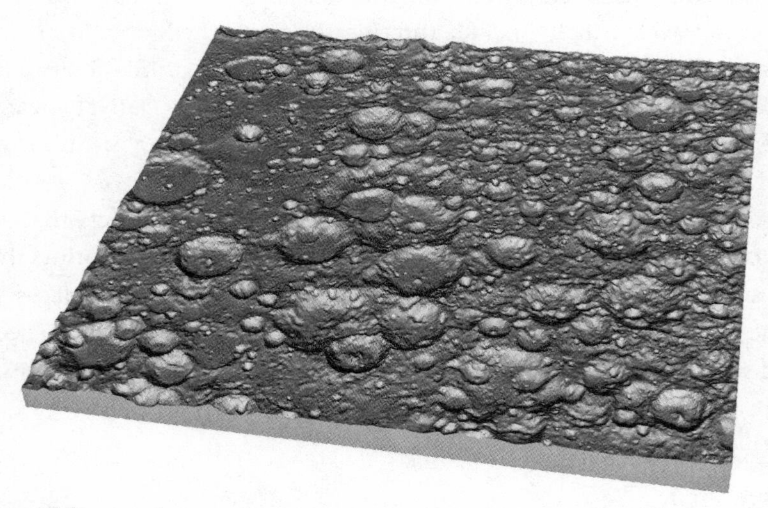

Figure 7.2
The Moon's far side is the hemisphere we can never fully see from our position on Earth due to the gravity has locked our Moon to always show the same side to our planet. It is very rugged in terrain, pockmarked and crowded with craters of all sizes. Until recently, the various space missions to the Moon had only landed on the near side, but in 2019 the first spacecraft (China's Chang'e 4) successfully landed on the far side.

NEARSIDE AND FARSIDE

These two 3D prints of the two sides of our Moon help show the differences between the near and far side. The near side is smoother due to large lava flows that filled in many craters billions of years ago. The far side is much more rugged in terrain, pockmarked with craters, including one of the largest craters in our entire Solar System (known as the South Pole–Aitken basin). Printed at their default size of 3 inches (8 centimeters) across these models cover an area approximately 250 miles (400 kilometers) on a side at a scale of five million to one.

TYCHO

Tycho is one of the most significant Moon craters, with particularly steep and sharp features because the crater is only about 110 million years

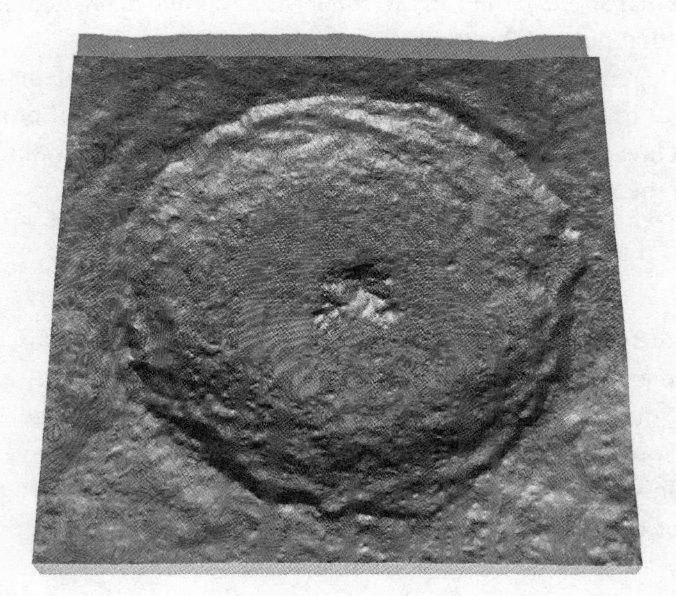

Figure 7.3

A crater on our Moon's near side that appears as a bright spot with rays of material stretching across the landscape, Tycho can be viewed with a backyard telescope. It is one of thousands of craters of such size at about 53 miles (85 kilometers) across but is a relatively young stamp on the lunar surface having formed from a meteor impact a little over a hundred million years ago.

old—quite young by Solar System standards—so there has not been much time for the edges to erode. A very popular target for backyard astronomers to locate with their telescopes, Tycho is about 51 miles (82 kilometers) in diameter. The summit of the central peak is 1.24 miles (2 kilometers) above the crater floor. The distance from Tycho's floor to its rim is about 2.9 miles (4.7 kilometers).

APOLLO 11 LANDING SITE

When Neil Armstrong took his famous first steps onto the lunar surface, he kicked around the soil and declared it was "fine and powdery." The lunar rocks at this spot helped reveal the Moon's fiery past to humans for the first time. The samples showed that the Apollo 11 landing site in Mare Tranquillitatis was once the site of volcanic activity, and the flat surface that afforded such an incredible vista was due to broad, thin flows of lava that flooded the region.

The 3D map of the area around the Apollo 11 landing site represents about 18 miles (30 kilometers) in each direction. The height of features has been exaggerated sixty times in the 3D model for better differentiation in the 3D print.

MARS

The fourth planet from the Sun has long been a source of fascination for humans. The twenty-first century may be when human explorers finally join their robotic counterparts on what may be a future destination of our species. Space agencies from around the world have joined NASA in sending probes and orbiters to the Red Planet and we can use some of these data to explore Mars as if humans were already able to walk on its surface.

GALE CRATER

When NASA scientists were deciding where the Curiosity rover should land on Mars, they considered many sites across the planet. They chose Gale Crater because this region, where a meteor struck about 3.5 billion

Figure 7.4
The site on the Moon where the famous Apollo 11 mission landed was once volcanic and is situated in Mare Tranquillitatis. The 3D representation shows about 18 miles (30 kilometers) in each direction of the landing site, with the height of features exaggerated about sixty times.

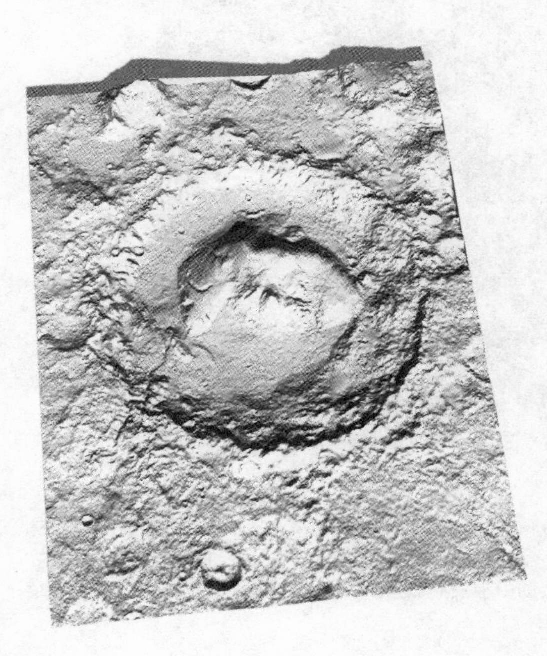

Figure 7.5
In this topographic representation, created with information gathered by the European Space Agency's Mars Express mission, the height of Gale Crater's features are enhanced by 300 percent. Since it arrived on Mars in 2012, NASA's Curiosity rover has been exploring Gale Crater, which is thought to be a dry lake and has a mountain in its center measuring about 18,000 feet (5.5 kilometers) high.

years ago, has many signs that liquid water was once present. This topographic model of Gale Crater shows the features discovered by ESA's Mars Express mission, which mapped the surface of the Red Planet. To see all the details, the height of the vertical structures has been enhanced by 300 percent. Printed at its default size of 4.3 inches (11 centimeters) across, this model covers an area approximately 125 miles (200 kilometers) across at a scale of 1.9 million to one.

PAHRUMP HILLS

Pahrump Hills is an outcrop at the base of Mount Sharp on Gale Crater, a peak at the center of the crater that stretches about 3 miles (5 kilometers) into the Martian sky. The Opportunity rover spent time in this flat slab and used its tools to grind a sample from the rock on its way to Mount Sharp in 2014. The region contains sedimentary rocks that scientists believe formed in the presence of water.

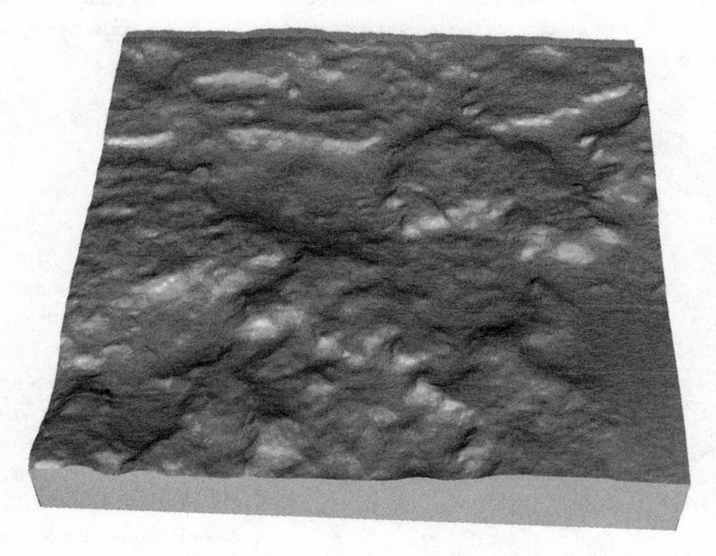

Figure 7.6

Part of Gale Crater on Mars, Pahrump Hills is a "mudstone outcrop" on the lower part of Mount Sharp, which has been explored by NASA's Curiosity rover. The rover has been drilling for samples on Pahrump Hills for study and analysis to learn about the planet's geologic history.

BLOCK ISLAND

What is it like to go rock hunting on another planet? Since landing on Mars in 2004, NASA's Opportunity rover has explored Mars, using its suite of scientific instruments to probe the geology it encounters. Composition measurements by Opportunity rover of this rock on the Martian surface indicates it is not from Mars at all, but an iron-nickel meteorite, the largest ever discovered on the Red Planet. Researchers have informally named the meteorite "Block Island." The meteorite spans about 2 feet (0.6 meters), approximately twice the size of the previous record-holding meteorite discovered on Mars.

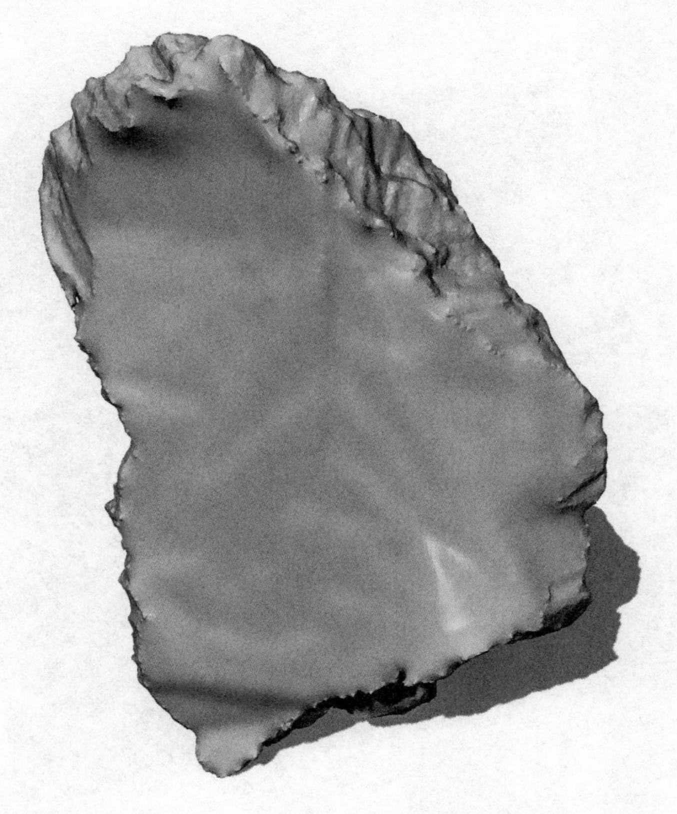

Figure 7.7

Block Island is a relatively large meteorite found on Mars. Measuring about 2 feet (0.6 meters) across and made of mostly iron and nickel, it is about twice the size of other similar objects in the surrounding area of the Red Planet.

VALLES MARINERIS

Valles Marineris, or Mariner Valley, is a vast canyon system that runs along the equator of Mars. The scale of Valles Marineris is enormous: it stretches five times longer than the Grand Canyon and is four times as deep. This topographic model represents a portion of Valles Marineris where the vertical features are again exaggerated for better visibility. The data come from NASA's Mars Global Surveyor, an orbiting spacecraft that has been collecting data on the Mars surface since 1999. Printed at its default size of 3 inches (8 centimeters) across, this model covers an area approximately 600 miles (1,000 kilometers) across at a scale of 12.5 million to one.

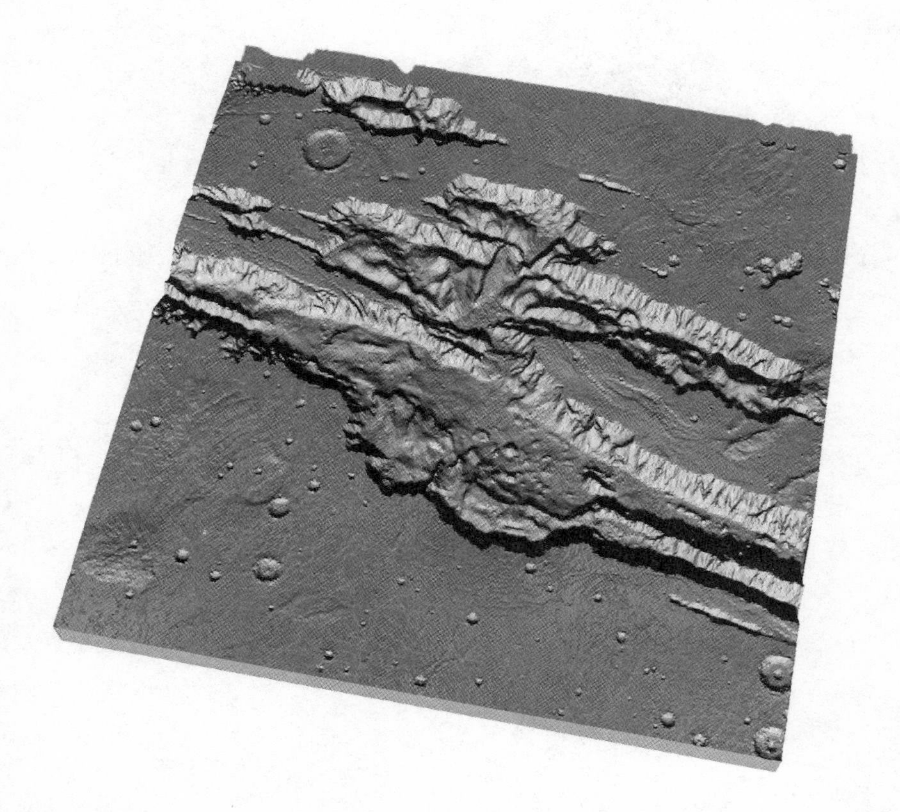

Figure 7.8
An enormous canyon mapped by NASA's Mars Global Surveyor, Valles Marineris stretches five times longer and four times deeper than the Grand Canyon. The vertical features in this topographical map are exaggerated to improve print quality.

AMBLING AROUND ON MARS

Some scientists sought better ways to interact with the enormous amount of data about Mars sent back to Earth from various rovers and orbiters. For example, NASA engineers have imported the hundreds of thousands of images taken by the Curiosity rover into a 3D model. This allows scientists (and the rest of us) to explore the Mars landscape in ways never before possible, using only a smartphone, computer, or VR goggles. Wander across the Martian dunes and valleys, zoom into interesting geological features like rock outcrops and mud cracks, or simply enjoy a "stroll" through Mars. This immersive web experience is free and was developed by the NASA Jet Propulsion Laboratory in collaboration with Google.

ASTEROIDS

About half of the mass of the Solar System's asteroid belt comes from four big asteroids that are each more than 200 miles (322 kilometers) wide. The rest is made up of smaller rocks of various sizes. While we await the day when humans can go to these bodies in person, we can use 3D data and technology to visit them virtually today.

VESTA

Vesta is the second most massive body in the asteroid belt that lies between Mars and Jupiter (Ceres being the largest). With a diameter of about 325 miles (525 kilometers), Vesta stretches for approximately the

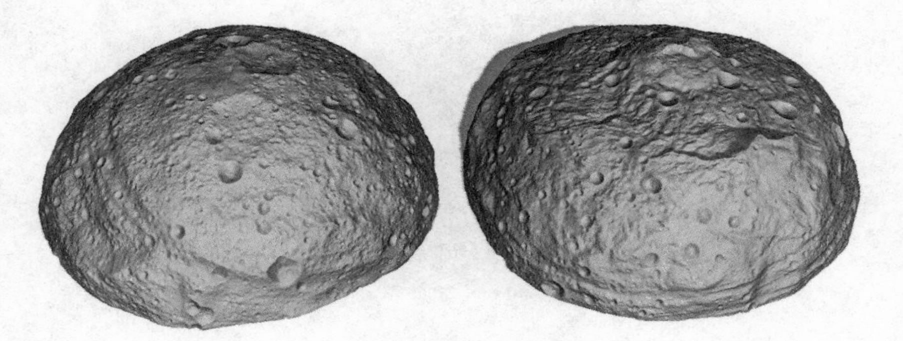

Figure 7.9
One of the largest asteroids in our Solar System, Vesta is located between Mars and Jupiter. This pockmarked and potato-shaped asteroid has a diameter of about 325 miles (525 kilometers), with a dramatic crater in its southern region.

distance between Pittsburgh and Philadelphia. In 2011, NASA's Dawn spacecraft arrived at this asteroid after a 4.3 billion-mile (6.9 billion-kilometer) journey from Earth. Scientists think that Vesta lost about 1 percent of its mass when it collided with another object. This impact created the Rheasilvia crater located in the asteroid's southern polar region. When this 3D model is printed at its standard size of about 5 inches (12 centimeters) wide, the scale is approximately four million to one. There are two halves that need to be printed to make one complete 3D representation of Vesta.

BENNU

Just over two decades ago in September 1999, astronomers discovered the asteroid that became known as Bennu. NASA then developed and launched a mission to visit the asteroid and collect a sample that will be

Figure 7.10

Scientists think the asteroid Bennu was formed about 4.5 billion years ago, making it an interesting age for researchers. This is one of the reasons NASA sent the OSIRIS-REx mission in 2020 to collect a sample of this asteroid 1,610 feet (490 meters) in diameter.

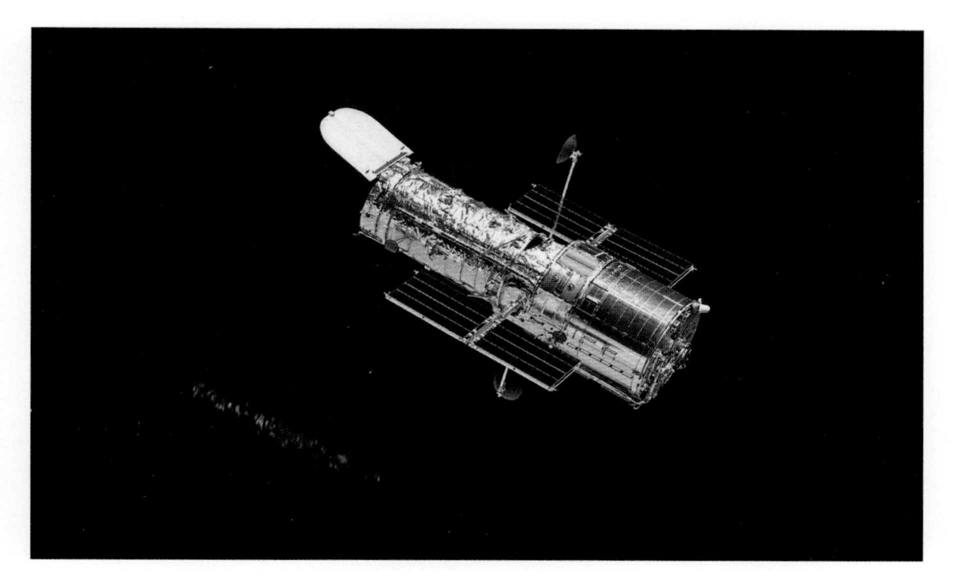

Plate 1

The Hubble Space Telescope celebrated its thirtieth year in orbit in 2020. Hubble has a unique design and is close enough to Earth that astronauts could repair and upgrade it with new technologies while the Space Shuttle program was still in operation. Hubble is one of NASA's longest-living and most valuable space-based observatories.

Plate 2

This artist's illustration depicts NASA's Chandra X-ray Observatory that was launched into space in July 1999. Chandra is about the size of a school bus and travels approximately a third of the way to the Moon at its farthest distance from Earth. Chandra and Hubble are both part of NASA's "Great Observatories" program, which commissioned four large and powerful telescopes to look at different types of light.

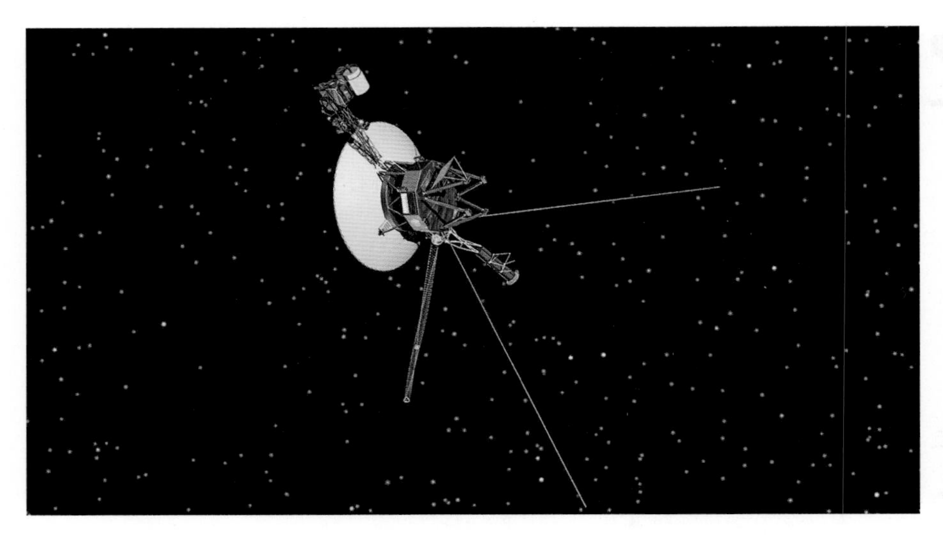

Plate 3
Voyagers 1 and 2 are identical twin spacecraft launched by NASA in 1977. The pair are still going over four decades later. In addition to their primary targets of Jupiter and Saturn, the Voyagers have gathered data on Uranus and Neptune along with forty-eight of the moons of planets in the outer Solar System.

Plate 4
This photograph shows the launch of the Space Shuttle Discovery on August 28, 2009. The Space Shuttle program was NASA's primary means to transport people and cargo into space from its beginning in 1981 through its end in 2011. There were five Shuttles that performed 135 missions, including two that were lost (Challenger and Columbia).

Plate 5

Launched in 1997, the Cassini-Huygens mission spent seven years traveling to the Saturn system. Once there, the Cassini orbiter circled Saturn for thirteen years collecting data. On January 4, 2005, the Huygens probe plunged to the surface of Titan, one of Saturn's most intriguing moons. This artist's impression shows the spacecraft against the backdrop of the rings of Saturn.

Plate 6

The Curiosity rover arrived on Mars on August 5, 2012, and has been looking for evidence of whether the Red Planet ever had the right conditions to support microbial life. This NASA mission is a car-sized machine with a seven-foot-long arm, ten scientific instruments, and seventeen cameras to perform its investigations.

Plate 7

Launched into space in 2007, Kepler was one of the first NASA telescopes dedicated to hunting for planets outside our Solar System (known as "exoplanets"). Over its lifetime, the mission identified 2,600 exoplanets by looking for the tiny dimming of light when a possible planet passed in front of a star (illustration).

Plate 8

As tall as a thirty-six-story building, the Saturn V was used in the Apollo missions of the 1960s and early 1970s as well as for launching the Skylab space station in 1973. In total, it flew thirteen flights including ten with people on board. Developed at the height of the Space Race with the Soviet Union, the Saturn V went from design to flight in just six years.

Plate 9

The vehicle that delivered Neil Armstrong and Buzz Aldrin to the surface of the Moon in 1969 was the Lunar Module. This machine had an unusual, almost bug-like shape. Its four legs were spread apart and equipped with big "feet" so it wouldn't sink on what was then an unknown lunar surface.

Plate 10

Our Moon is about a quarter of the size of the Earth and, at a distance of about 240,000 miles (380,000 kilometers), the nearest neighbor to us. Interesting in its own right, the Moon has important effects on Earth including helping produce tides in our oceans through gravity. The Moon is tidally locked to the Earth, meaning the same side always faces our planet.

Plate 11

This feature in the southern hemisphere of the Moon is known as Tycho Crater. It is not the biggest crater on the Moon with a diameter of 53 miles (85 kilometers) across, but it may be one of the youngest. One sign of its relative youth are the rays, visible as bright streaks, left over from material ejected during the impact event just over a hundred million years ago.

Plate 12

Despite the commonly used phrase, there is no "dark side" of the Moon. Rather, there is a side that perpetually faces away from Earth known as its "far side." While the far side does receive its share of sunlight, it is different from the nearside that we see from Earth's surface. For example, there are significantly fewer craters on this hemisphere on the Moon.

Plate 13

This image from the Lunar Reconnaissance Orbiter released in 2011 shows the landing site of Apollo 17, the final Moon mission of the Apollo program. The dark splotch with the bright middle near the center of the image is the Challenger lunar module landing site, while the ribbon-like features are human footpaths and tracks from a lunar rover.

Plate 14

Mars is the fourth planet from the Sun and a frequent focus of both science fiction and scientific exploration. Since the Viking missions in the 1970s, humanity has sent robotic explorers to investigate the Red Planet—including looking for signs of liquid water and microbial life. The planet gets its color from iron oxide that is found across its surface.

Plate 15

On Mars, there are many fascinating geological features. These sedimentary layers in a large canyon on Mars were imaged by the High Resolution Imaging Science Experiment (HiRISE) aboard NASA's Mars Reconnaissance Orbiter. This image spans just 0.6 miles (1 kilometer) across.

Plate 16
This rock on Mars, nicknamed "Block Island," is not actually from that planet. Rather, scientists think it is a meteorite composed mainly of iron-nickel, a relic from the formation of the Solar System. This image was taken by the navigation camera aboard the Opportunity rover on July 28, 1999. The meteorite spans about 2 feet (60 centimeters) across.

Plate 17
Vesta is the second largest asteroid (behind Ceres) in the belt between Mars and Jupiter. For a sense of scale, the diameter of Vesta would stretch about the distance between Pittsburgh and Philadelphia. NASA's Dawn mission captured this image of Vesta right before it departed its orbit on August 26, 2012.

Plate 18

The Origins, Spectral Interpretation, Resource Identification, Security-Regolith Explorer is better known by its acronym, OSIRIS-REx. This mission traveled to the asteroid Bennu, seen in this image, and collected a sample of rocks and material from its surface in 2020. It is scheduled to return that sample to Earth in 2023.

Plate 19

This image of our home planet, Earth, was taken on July 6, 2015, by the Deep Space Climate Observatory (nicknamed DSCOVR) from 1 million miles (600,000 kilometers) away. There is a history of images of Earth taken from space, including perhaps the most famous one known as the "Blue Marble" taken by the Apollo 17 astronauts on their way to the Moon.

Plate 20

As Hurricane Katrina bore down on the southeastern coast of the United States on August 28, 2005, an instrument aboard NASA's Terra satellite captured this image of the monster storm. With winds of up to 160 miles per hour (257 kilometers per hour), Katrina was one of the most powerful hurricanes ever to strike the country.

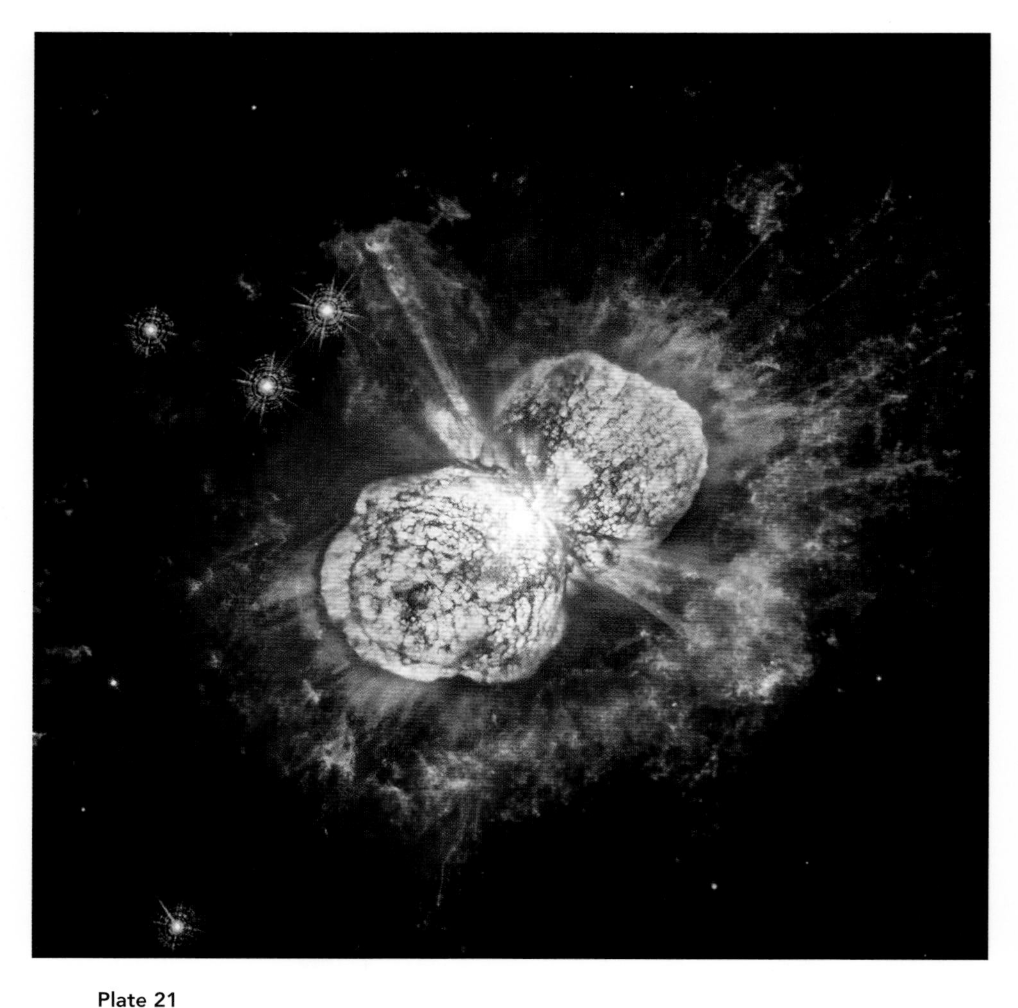

Plate 21

Astronomers have been watching Eta Carinae since it underwent the "Great Eruption" in the 1840s, temporarily becoming the second-brightest star in the sky. Eta Carinae is a double star system and its violent outbursts have created these two giant lobes of material seen in this image from the Hubble Space Telescope. One day, Eta Carinae may explode as the next supernova in the Milky Way.

Plate 22

The Eagle Nebula, also known as Messier 16, is the site of the spectacular star-forming region that astronomers have dubbed the "Pillars of Creation." The tall pillars of gas and dust house stellar nurseries, where new stars are being created. This region became famous when the original Hubble image was released in 1995.

Plate 23
NGC 1555, also known as Hind's Variable Nebula, is a cloud of gas and dust. It is being illuminated by the star T Tauri, which was discovered by John Russell Hind in 1852. T Tauri went on to become the prototype for the entire class of young stars that share its name. This original T Tauri star is located about 460 light-years from Earth.

Plate 24
Westerlund 2 is a cluster of young stars—about one to two million years old—located approximately 20,000 light-years from Earth. Data in visible light from Hubble reveal thick clouds where the stars are forming. Westerlund 2 contains some of the hottest, brightest, and massive stars in the Milky Way galaxy.

Plate 25

Look just below the middle of the three stars of the belt in the constellation of Orion to find the Orion Nebula, which can be seen without a telescope. With a powerful telescope like Hubble, the view is much different. More than three thousand stars of various sizes appear in this image. The bright central region is the home of the four heftiest stars in the nebula. The stars are called the Trapezium because they are arranged in a trapezoid pattern.

Plate 26
The Small Magellanic Cloud (SMC) is one of the Milky Way's closest galactic neighbors. Even though it is small, the SMC is so bright that it is visible to the unaided eye from the Southern Hemisphere and near the equator. This Hubble image shows glorious details of one area of star formation within the SMC known as NGC 602, giving astronomers more information about stars that are born and evolve.

Plate 27
In 1054 A.D., observers in several countries reported the discovery of a "new star" in the constellation of Taurus. Today, this object is known as the Crab Nebula and astronomers know it is powered by a rapidly rotating dense object called a neutron star that was formed when a massive star ran out of fuel and collapsed on itself. This image contains X-ray, optical, and infrared light.

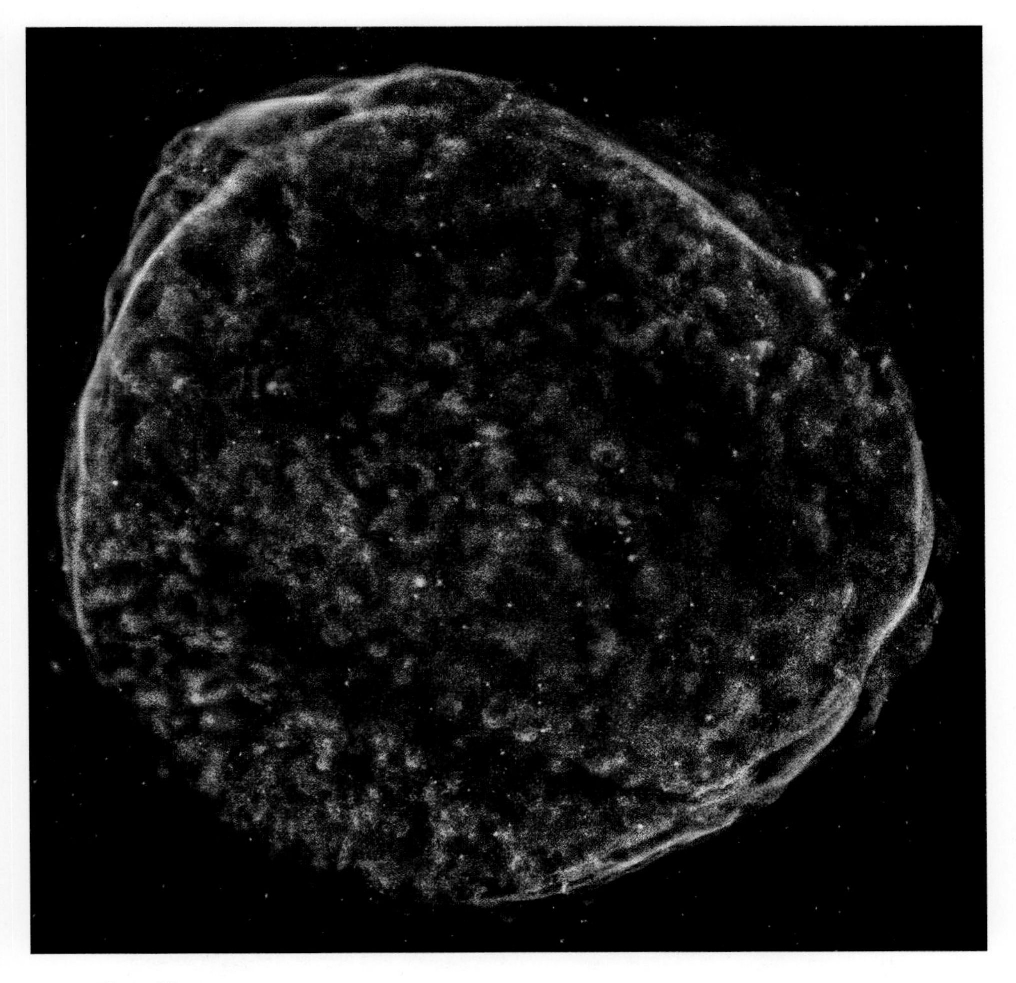

Plate 28

Skywatchers over a thousand years ago marked the appearance of this object on May 1, 1006. It was brighter in the night sky than Venus for a period and visible during the day for more than two weeks. This X-ray image from Chandra X-ray Observatory shows the glowing tapestry that is a debris field left behind after a star exploded. Today we call this Supernova 1006.

Plate 29

The Earth lies about 26,000 light-years from the center of the Milky Way. Between us and the supermassive black hole that resides there are great amounts of dust and gas that obscure the view. Some types of light, including X-rays, can penetrate this veil. This image from Chandra X-ray Observatory shows the region around our Galaxy's black hole, called Sagittarius A*, which is found in the middle of the bright white source in the center.

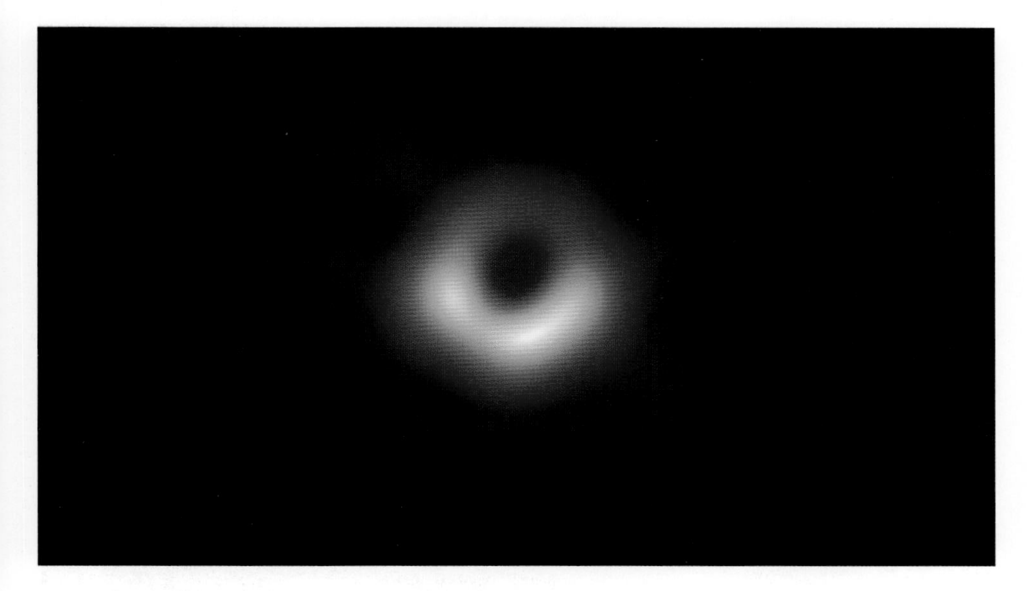

Plate 30

M87 is an elliptical galaxy in the Virgo galaxy cluster, around sixty million light-years away from Earth. For years, scientists have known that a supermassive black hole weighing several billion times the mass of the Sun sits at the center of M87. This dark portrait—like a silhouette—of the event horizon was obtained of the supermassive black hole by the Event Horizon Telescope (EHT), an international collaboration whose support includes the National Science Foundation.

Plate 31

This stunning image of the central region of the Milky Way combines three kinds of light. Near-infrared light from Hubble (yellow) outlines energetic regions where stars are being born. Infrared data from Spitzer (red) show glowing clouds of dust containing complex structures. X-rays from Chandra (blue and violet) reveal gas heated to millions of degrees by stellar explosion and outflows from the Galaxy's supermassive black hole.

Plate 32

The galaxy is officially named Messier 51 (M51) or NGC 5194, but often goes by its nickname "Whirlpool Galaxy." Like the Milky Way, the Whirlpool is a spiral galaxy with spectacular arms of stars and dust. Unlike our Galaxy, M51 has a smaller companion galaxy that has started to merge with it. This image shows M51 is visible light from Hubble (red, green, and blue) and X-rays from Chandra (purple).

Plate 33

M100 is beautiful spiral galaxy, a rotating system of stars, gas, and dust, similar to our own Milky Way. This image from the Hubble Space Telescope reveals the galaxy's prominent spiral arms and individual stars. A burst of star formation is occurring in the dusty lanes of the arms where numerous stellar-mass black holes—those about five to ten times the mass of the Sun—also reside.

Plate 34

Messier 86 (M86) is generally classified a lenticular galaxy (a hybrid between an elliptical galaxy and a spiral one). It is one of the brightest members of the Virgo galaxy cluster, a collection of 1,300 galaxies. This image of M86 was captured by the Sloan Digital Sky Survey, a wide-field optical telescope in New Mexico.

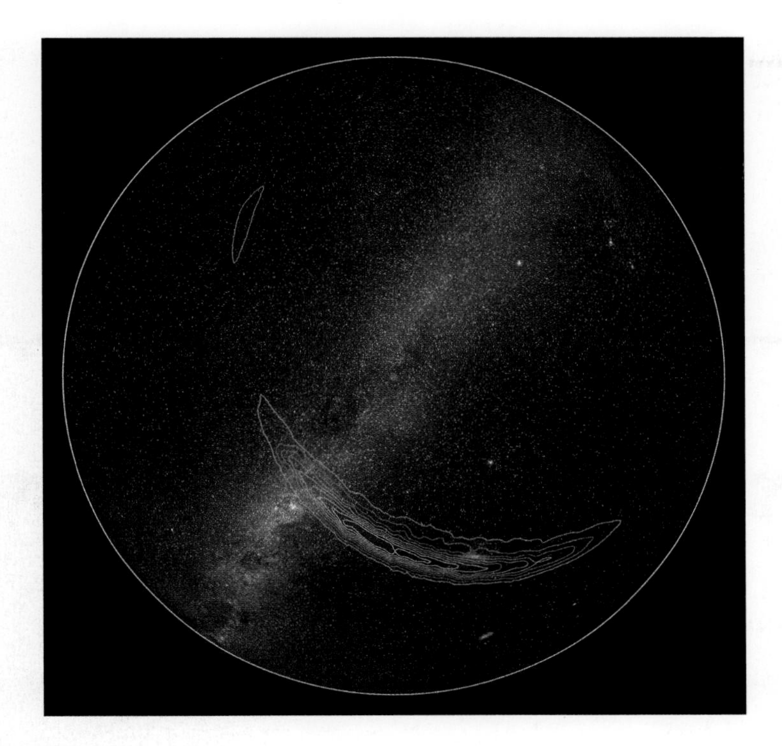

Plate 35

This graphic shows the approximate location of the source of gravitational waves detected on September 14, 2015, by the twin Laser Interferometer Gravitational-Wave Observatory (LIGO) facilities' overlaid sky map of the Southern Hemisphere. The colored lines represent different probabilities for where the signal originated, with purple being the highest. This historic discovery marked the first time that scientists detected gravitational waves.

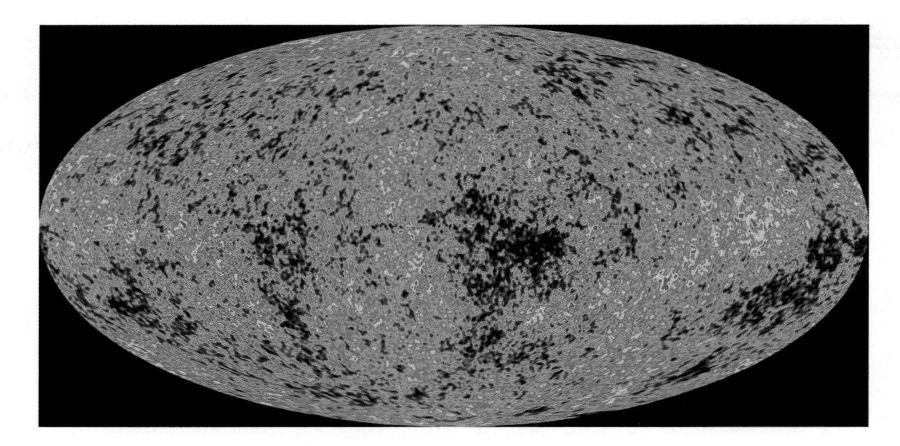

Plate 36
Scientists have been able to use specialized telescopes to measure tiny variations of temperature across the sky to help pin down the age of the Universe (13.8 billion years) as well important features such as its shape and composition. This signal, which is a relic from the Big Bang itself, is known as the "cosmic microwave background" (CMB). After its launch in 2001, the Wilkinson Microwave Anisotropy Probe (WMAP) collected extremely high-precision data of the CMB. This WMAP map shows differences in color that represent fluctuations in the CMB from about 400,000 years after the Big Bang.

Plate 37
The Cassiopeia constellation is formed by five stars that are bright due to their relative proximity to Earth, all within a few hundred light-years to Earth, so nearby in astronomical terms. This group of stars is visible all year long in the Northern Hemisphere and in the northern part of the Southern Hemisphere during spring.

Plate 38

The constellation of Ursa Major (or "Great Bear") is perhaps most famous for an asterism within it. Seven stars create the Big Dipper, a prominent pattern or group of stars that is not actually a constellation known as an asterism. Ursa Major and its asterism are significant in many cultures because of their prominence in the Northern Hemisphere. This is one reason it appears on the state flag of Alaska.

Plate 39

Sagittarius, the constellation, contains seventeen named stars, including seven that are rather bright and therefore relatively easy to spot with the unaided eye. It has a well-known asterism called the "Teapot" and the constellation encompasses the center of the Milky Way, which is why the black hole at the center of our Galaxy is named Sagittarius A*.

Plate 40

In Greek mythology, Centaurus is a centaur, a hybrid of man and horse, but even more ancient cultures such as the Babylonians also noted this star pattern. The Centaurus constellation has a large, four-sided shape that is supposed to represent the human head and torso, attached to two legs. Alpha Centaurus is the triple star system that contains Proxima Centauri, the closest star to our Sun, and forms one of the Centaur's feet.

returned to Earth in 2023. Bennu was formed when many rocks in the Solar System became compressed through gravity about 4.5 billion years ago. Because of this asteroid's advanced age, scientists are eager to literally get their hands on pieces of Bennu to learn more about the earliest days of our planetary neighborhood. Bennu is a much smaller asteroid than Vesta, with a diameter of about 1,610 feet (490 meters). This is roughly a bit more than once around a standard quarter-mile track.

EROS

The oblong-shaped body of Eros is important for it was the first asteroid to be orbited by a spacecraft in 1998 and then landed upon with the same spacecraft (NASA's Near Earth Asteroid Rendezvous, or NEAR, mission) three years later. Scientists also consider this asteroid called Eros to be the first "near-Earth" asteroid to be discovered in 1898. This is a class of Solar

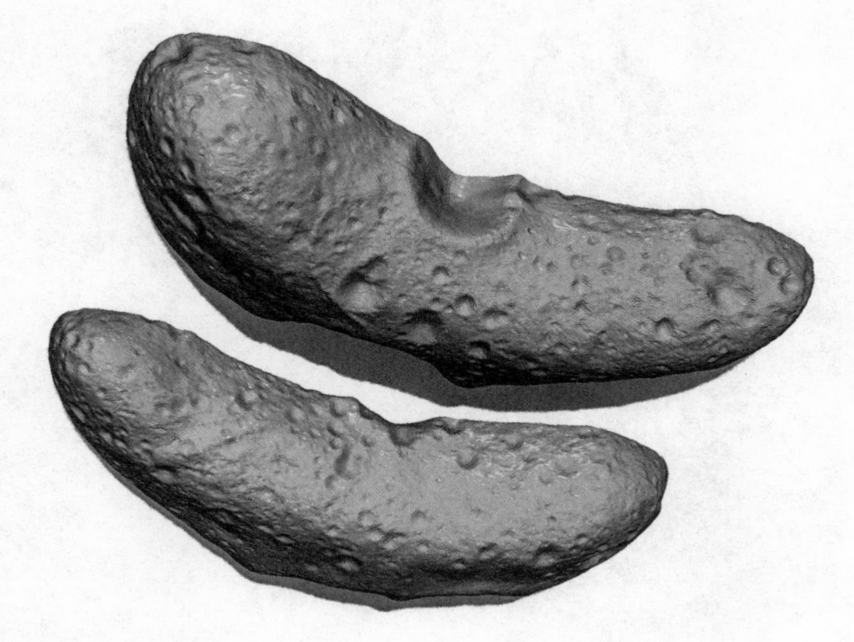

Figure 7.11

Eros, in the shape of a fingerling potato and about 20 miles (34 kilometers) in size, was the first asteroid to be orbited by a spacecraft. NASA's Near Earth Asteroid Rendezvous (NEAR) mission not only orbited the asteroid but also landed on it.

System bodies that have the potential to one day impact Earth, or at least come relatively close. In the case of Eros, that potential threat would not materialize for at least a million or two years from now. Eros has a length of about 20 miles (34 kilometers) and when it's printed at its default size the 3D model is about 2 inches (5 centimeters) long. Like Vesta, the 3D Eros needs to be printed as two halves to make a complete whole.

EARTH

As the most germane Solar System object to humanity, it seems like it would be remiss to not include some 3D examples of our home planet. But Earth is

Figure 7.12

Our own home planet is of continual interest to humans and there is still much to explore. This 3D map of the topography of Earth is greatly exaggerated to depict elevations of the land masses.

also the most familiar object, and its topographical features have been available through tactile globes and other products for decades. There are some elements within Earth and its complex atmosphere, however, that have been given new treatments with the advancements in 3D technologies.

HURRICANE KATRINA

One area where 3D imaging has progressed significantly is powerful storms such as hurricanes. An example is the 3D model made from visible and thermal intensities from Hurricane Katrina, which ravaged parts of the southeastern United States in 2015. When a National Ocean and

Figure 7.13
This 3D model of Hurricane Katrina was created from infrared and visible data captured on August 28, 2005, when Katrina had become a Category 5 storm in the Gulf of Mexico. Katrina's center was south-southeast of New Orleans, Louisiana, with hurricane force winds at 175 mph (280 km/h).

Atmospheric Administration (NOAA) satellite captured these data, Katrina was a Category 5 hurricane with sustained winds of about 175 miles per hour (280 kilometers per hour).

ELSEWHERE IN THE 3D SOLAR SYSTEM

There are ongoing attempts to create data-based 3D models of many other objects in our Solar System. From Venus to Jupiter, from Io to Pluto, data are being collected for all these celestial bodies. However, most of these 3D collections so far cover only a small, specific region of each body or do not necessarily contain enough detail to make an interesting 3D print. If you are curious about what else is available, visit this book's companion website at https://starsinyourhand.pubpub.org/ for some of the latest models of the 3D Solar System.

HOW FAR, HOW BIG

One of the challenging aspects about space is wrapping your head around the immense scale of objects in space and the distances between them within the cosmos Once we think about destinations beyond planet Earth, we quickly lose familiar reference points. However, it's worth trying to keep great sizes and distances in some sort of context.

Astronomers use the somewhat misleading term of "light-year" to keep track of distance. It's potentially confusing because the term suggests it's a unit of time, but it is, in fact, a measure of how far light travels in one year. At the conceptual maximum speed limit, according to Einstein and others, of about 670,000,000 miles per hour, the unit of a light-year translates to about 6,000,000,000,000 miles (or 10 trillion kilometers).

Throughout this book, we will use light-years to convey distance. Our Sun is about 93 million miles away from Earth, which makes it about eight light-seconds away (the distance light travels in eight seconds.) The next nearest star is about four light-years from Earth. The center of the Milky Way galaxy? About 26,000 light-years on our cosmic horizon. Once we consider objects outside our Galaxy, the distances covered include millions or even billions of light-years.

It's also important to keep track of how big these objects are. If you could slice through the middle of Earth in a straight line, you'd travel just about 7,900 miles (12,700 kilometers) across its diameter. But Earth is a small planet for our Solar System compared to, say, Jupiter, which has a diameter of about 88,800 miles (142,900 kilometers) through its equator. You need to take another step on the power of ten to get to the Sun's diameter at about 865,000 miles

(1.39 million kilometers). Astronomers estimate that our Milky Way galaxy is some 100,000 light-years across, or about 600,000,000,000,000,000 miles (1,000,000,000,000,000,000 kilometers) wide.

When we leave the Solar System, it often makes sense to return to light-years because things in space get amazingly large. For example, some supernova remnants can span trillions and trillions of miles so astronomers give the size across the sky in light-years. We will provide numbers and units throughout this book to give you a sense of the intellectual shrink ray we've used to allow you to spin these objects on your screen or hold one in your hand as a 3D print.

8

MUCH MORE THAN TWINKLING: STARS

Go out on a dark and clear night and, if you are lucky enough to be far away from artificial lights, you will see hundreds or even thousands of stars. Stars are some of the first acquaintances we make with space. For millennia, humans have looked at the stars, studied them, and woven them into our folklore, religion, and identity.

Our relationship with stars, however, goes even further beyond to a physical level. If we look at the list of elements that make life possible on Earth—oxygen, iron, calcium, and so on—we find that they are only made inside the nuclear furnaces of stars.

Just after the Big Bang, the Universe was a very hot and dense ocean of particles. After a few hundred thousand years, it cooled enough so that atoms could form. These atoms were mainly helium and hydrogen, the latter of which remains the most abundant element in the Universe today.

The first generation of stars formed about 150 million years after the Big Bang, and since then the successive generations of stars have come from basically the same process. Clouds of gas and dust are scattered throughout most galaxies. Sometimes, turbulence ripples through clouds of gas and dust, creating knots of material often in pairs or trios. When these nodules are dense enough, they begin to collapse under their own gravitational attraction.

Gravity continues on its path and pulls the cosmic building material of the star toward its center, while some of it encircles the developing star in a disk. Eventually, the center of the burgeoning star is compressed so much that nuclear fusion ignites.

The process of converting hydrogen and helium into other elements generates pressure outward, balancing the inward force of gravity. This yin/yang relationship allows the star to exist.

This is how our Sun—and billions and billions of other stars—formed. The Sun and the rest of the objects in our Solar System were born about five billion years ago, and astronomers estimate that our Sun will last in its current stage for another five billion years or so.

For the bigger picture of stellar evolution, what happens after stars form? The path a star takes is strongly driven by the amount of mass it has once it is born. Smaller stars smolder, in stellar terms, going through their fuel supplies slowly. This means that they can last for many billions of years. Eventually, these stars cool off and shrink, becoming cosmic embers.

On the other end of the spectrum, the largest stars burn brightly and end their lives much more quickly—over just millions of years. Some of the biggest stars last only a fraction as long compared to their less massive cousins and, once they run out of suitable atoms to fuse, explode as supernovas. These events are some of the most powerful in the Universe, sometimes briefly producing more light than an entire galaxy.

STELLAR EVOLUTION

There is a wide range of what constitutes a star. We can only look at stars (or any objects in space) as they appear to us now in the light that arrives at our doorstep. Like looking at photographic snapshots of many generations and branches of a family tree, this helps astronomers develop the full story of how the diverse population of stars are born, live, and die.

Smaller stars are typically cooler over remarkably long lifetimes before quietly petering out. Bigger stars like our Sun have more active but shorter lives with several dramatic phases before entering their stellar retirement. And the most massive stars often end violently, producing tremendous explosions as well as exotic objects such as neutron stars and black holes.

Each limb in the tree of stellar evolution can break off into many branches. We include some, but not all, of the "leaves" that astronomers have found that stars can grow into.

The illustrated chart (figure 8.0) outlines the major paths that stars can take over their lifetimes.

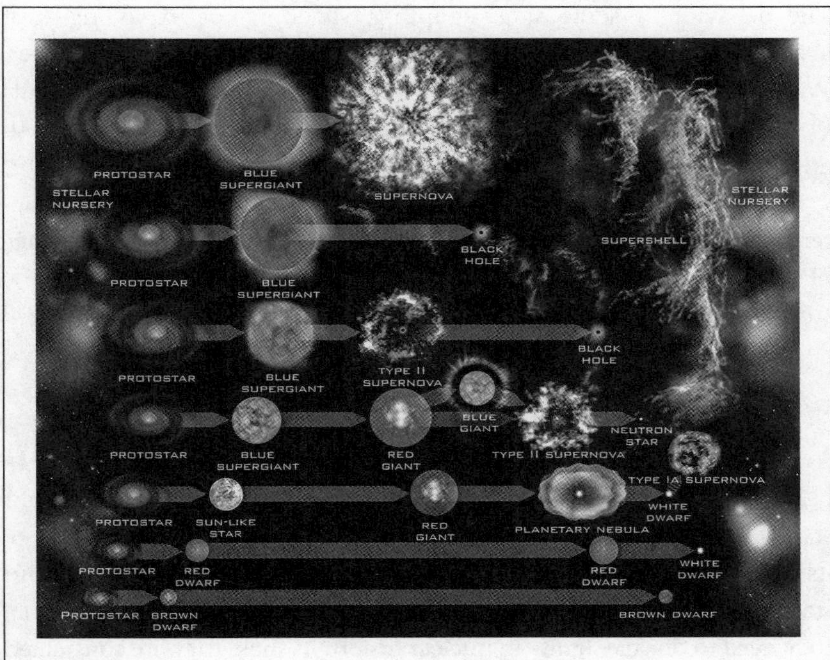

Figure 8.0
The fate of a star depends on its mass, which this illustrated chart outlines. Our Sun is of average size and middle-aged but is of course crucial to our existence. Other stars can be much smaller and live longer lives, or much more massive and live short, furious lives.

Supernova explosions act as super-seeders, dispersing the elements they have forged in their stellar furnaces out into space. From there, these elements can make their way into the giant clouds of gas and dust that eventually form the next generation of stars and the planets that may orbit around them.

Like every other star in the night sky, our Sun's cosmic gene pool was filled by the debris of *stars that came before the Sun*. We are directly linked to generations of stars that came before by the atoms and molecules inside our bodies and throughout our planet.

What is mesmerizing about stars is that practically all of them—even "average" ones like our Sun—are full of surprises. The winds thrown off by baby stars and the structures they create are three-dimensional assemblies.

The behavior of some stars—erupting, shedding, or exploding—cast shapes across space that can only be fully studied from all perspectives. With our current technologies, astronomers have captured large amounts of high-precision data of relatively nearby stars that allow us to experience these objects and what they are doing more fully.

The collection of stars in various states of evolution in 3D in this chapter represent just the tip of the proverbial iceberg in terms of the range of size, phase, and behavior of what we see as twinkling dots from the surface of Earth.

T TAURI

As they form, stars go through some intermediary steps on their journey to becoming a full-fledged star. A "T Tauri" star is a very young, lightweight star around ten million years old or less. (The name comes from the first of its kind that was identified in 1852 in the constellation of Taurus.) T Tauri stars are still in the process of collecting and condensing all the material they need to one day ignite in nuclear fusion. As such, they are considered a middle step between a still-growing star and a fully-formed star like the Sun. T Tauri stars often have relatively large disks of material around their equatorial regions, a remnant of the star formation process.

This 3D file, which was built from mathematical modeling, shows not only the young star, but also several features. These include a disk of gas and dust that accumulates around the star's center of gravitational attraction (called an "accretion disk"); its magnetic field; looping structures made of blisteringly hot, electrically charged gas; and streams of disk material disrupted by the magnetic field and being drawn back to the center of the star. These are all important components of as the T Tauri star matures into stellar adulthood.

ETA CARINAE

Astronomers have kept a close eye on the double star system Eta Carinae since the middle of the nineteenth century when it temporarily became the second brightest star in the sky. Modern telescopes revealed that this event, dubbed the "Great Eruption," unleashed as much as ten times the

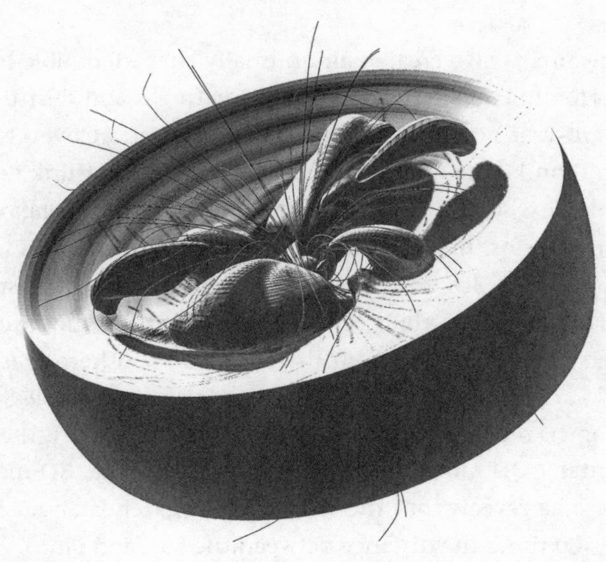

Figure 8.1
T Tauri stars are very young stars—about ten million years old—that have about the mass of the Sun and represent an early stage in stellar evolution. This 3D view of a T Tauri star shows its disk of gas and dust, its magnetic field, and streams of disk material disrupted by the magnetic field and being drawn back to the center of the star.

Figure 8.2
Eta Carinae is a double star system and could potentially be the next supernova in our home galaxy. The model depicts its double-lobed shell that's called the Homunculus Nebula, showing the protrusions, trenches, holes, and other irregularities throughout the gaseous material.

mass of the Sun. It also created an unusually shaped, double-lobed shell called the Homunculus Nebula. It is filled with gas and dust that is now about a light-year across and continues to expand at over 1.5 million miles (2 million kilometers) per hour. Some scientists think Eta Carinae could be the next supernova to explode in the Milky Way galaxy.

With protrusions, trenches, holes, and other irregularities in the gaseous material, this 3D print reveals the texture of what astronomers think the Homunculus Nebula is like. To get this three-dimensional view, astronomers combined data from the European Southern Observatory's (ESO) Very Large Telescope (VLT) with data from the Hubble Space Telescope. Using 3D modeling software designed for astronomy, the scientists built a printable 3D model of the Homunculus Nebula. 3D modeling of the Eta Carinae reveals that the Homunculus extends about five light-years, or 7,500 times the distance between the Sun and Pluto.

MESSIER 16

The Eagle Nebula, also known as Messier 16, contains the spectacular star-forming region known as the "Pillars of Creation." Three giant columns (which to some resemble buttes found in the American Southwest) of cold gas bathed in the scorching light from a cluster of young, massive stars. The Hubble Space Telescope made the Pillars of Creation one of the most famous objects in space after it released the first images of it in 1995.

Astronomers combined data from the Hubble and Very Large Telescope in Chile to peer around the Pillars in this 3D rendering and understand how each one is oriented to the other in space. The pooled data from space and the ground showed that each pillar is separated from each other in space. Looking at this system in 3D reveals that the tallest pillar may be pointed toward Earth while the other two are leaning away. This, plus an unrelenting bombardment of radiation from young stars in the vicinity, may explain why the tip of the tallest pillar appears to be the brightest.

FAST RADIO BURSTS

In 2007, scientists announced the discovery of something unusual: flashes of light that give off more energy than the Sun does over a hundred years.

Figure 8.3
Messier 16 contains the famous region known as the "Pillars of Creation," due to its interesting shape. An important aspect of the 3D model is that it demonstrates how the pillars of gas and dust are separated from each other, and that the tallest pillar seems to be pointed toward Earth while the other pillars are pointed away from Earth.

These flashes, however, last only a fraction of a second. Because they are so quick, astronomers had a hard time homing in on where they were located and what exactly they were. The scientific jury is still out about these mysterious surges, some of which behave differently from others, but astronomers categorize them as "fast radio bursts" or "FRBs."

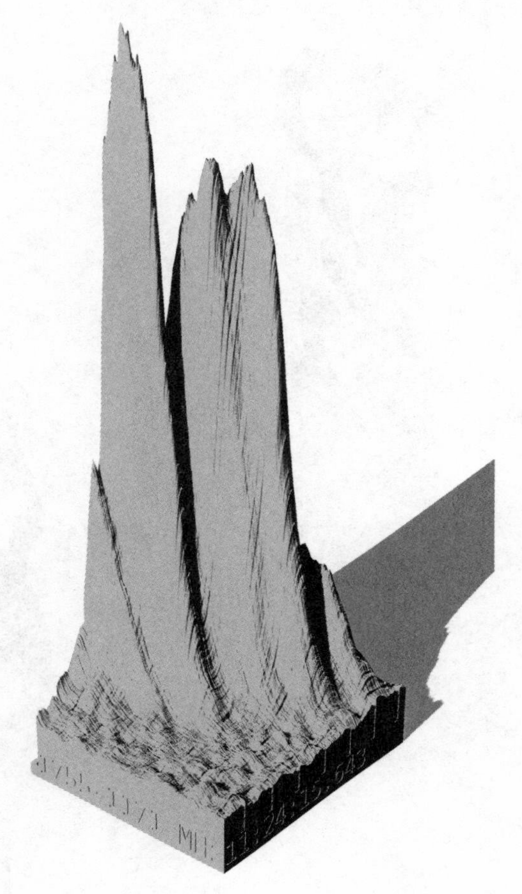

Figure 8.4
Fast Radio Bursts (or FRBs) are cosmic flashes that have been observed by radio telescopes all over the world, but last for just a fraction of a second. Repeated detections of such outbursts from FRB 121102 were translated into this 3D model to show how radio emission changes over time.

Astronomers have used radio telescopes in different parts of the world to observe that the bursts from one FRB, known as FRB 121102, repeat—but not at regular intervals. Every so often, there will be an hour or two when many bursts are detected that begin to decline in intensity over the flare up. This 3D model shows numerous individual bursts over time from FRB 121102. The frequency, or wavelength, in radio waves of each burst is represented by how high each peak is, while the horizontal shows the time interval in milliseconds (thousandths of seconds).

WR98A

Stars that are older and have mass that is forty or more times greater than the Sun's can become what astronomers refer to as "Wolf-Rayet" stars (named after two of their discoverers). In this stage of stellar life, the outer layer of the star has been driven off, exposing the core that is largely composed of helium. For some stars, the Wolf-Rayet phase is an intermediary step before its core becomes a white dwarf; for others, it happens before the star explodes as a supernova. Astronomers think that Wolf-Rayet stars

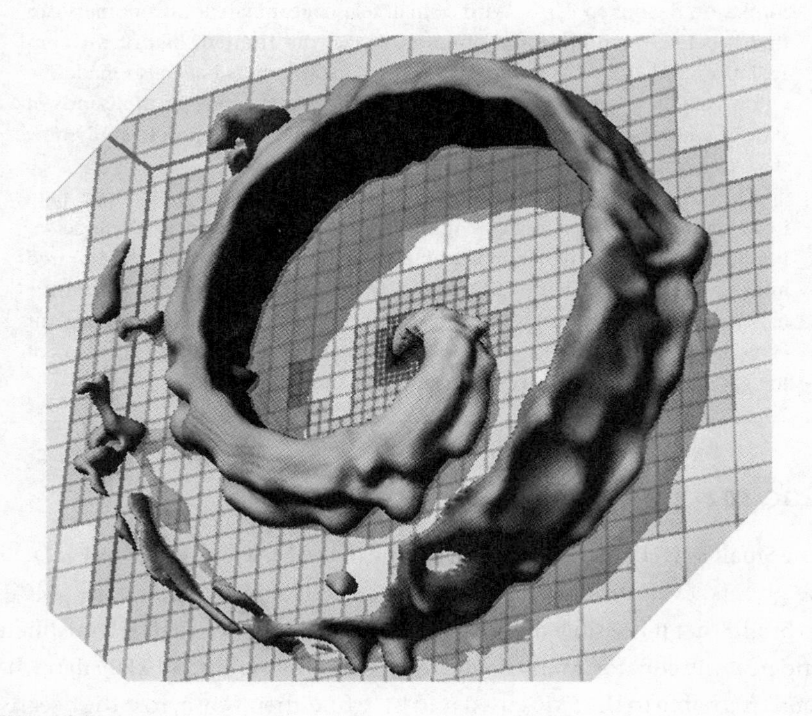

Figure 8.5
WR98a is a system that contains multiple stars and a collection of cosmic dust. This simulation was created from a mixture of predicted observations and collected data that show the spiral or pinwheel-like structure of the gas and dust around the system.

are the descendants of very massive stars that once possessed over twenty times the mass of the Sun.

WR98a is a system that contains a Wolf-Rayet star in orbit around another star as well as a surprising collection of dust. (The presence of dust in these systems is surprising because these giant stars blast out so much radiation that astronomers expect they would destroy any dust grains in their vicinity.) Astronomers used 3D modeling in combination with infrared observations to understand how the system produces a spiral tail of outflowing gas and dust.

FLYING THROUGH THE ORION NEBULA

At a distance of 1,500 light-years, the Orion Nebula is one of the closest star formation regions to Earth. With a small telescope, amateur astronomers can find it below the middle and dimmest star in the Hunter's belt of the constellation that shares its name. Professional astronomers have also made the Orion Nebula a favorite target. The Orion Nebula is home to thousands of stars of various sizes and stages, making it an excellent location to study how stars are born and behave during their stellar childhood.

This unique 3D fly-through of the Orion Nebula was assembled from data from multiple NASA telescopes in space. Researchers built a 3D model of the known topography of the region and then overlaid optical and infrared images over the 3D "frame." Ultraviolet radiation and stellar winds from the massive central stars carved out an enormous cavity in the wall of a giant cloud laced with dust. The viewer goes on a journey through the contours of the gas and dust to see deeply embedded stars in the cluster.

NGC 602

The Small Magellanic Cloud (SMC) is one of the Milky Way's closest galactic neighbors. Even though it is a small, or so-called dwarf, galaxy the SMC is so bright that it is visible to the unaided eye from the Southern Hemisphere and near the equator. Many navigators, including Ferdinand Magellan who lends his name to the SMC, used it to help find their way across the oceans. Within the SMC lies a bright young cluster of stars that astronomers call NGC 602, which they study for many reasons including that it contains a collection of similar stars to our Sun outside the Milky Way.

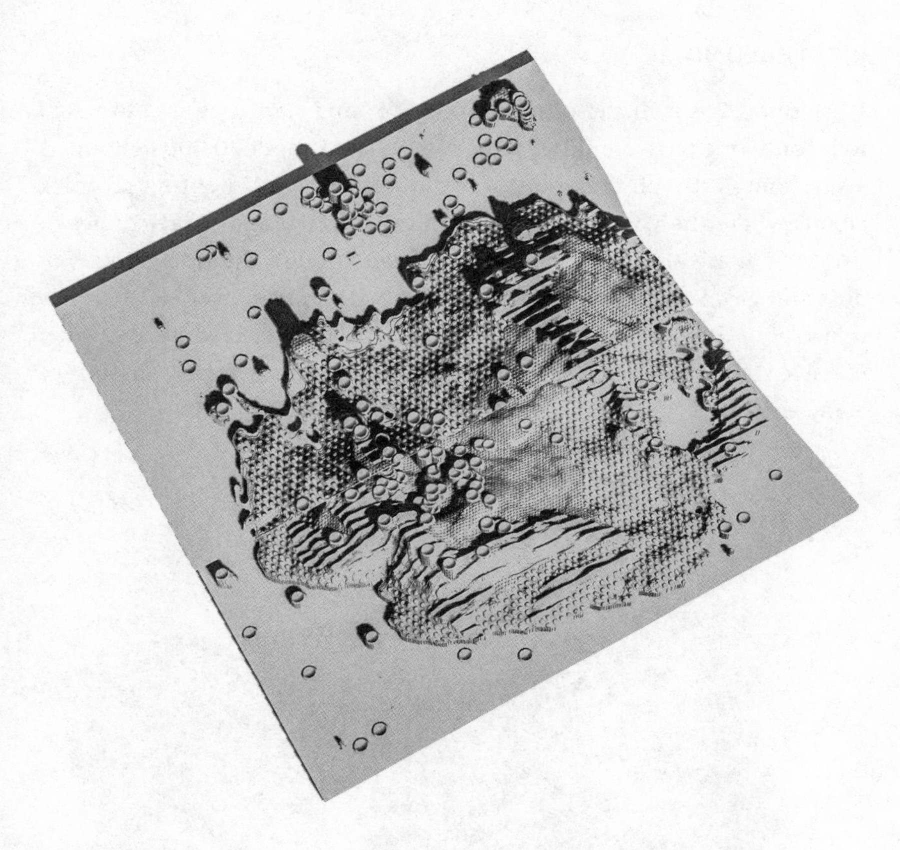

Figure 8.6
The star cluster NGC 602 is in the Small Magellanic Cloud, a dwarf galaxy close to our own Milky Way. Young stars that are only about four million years old are located primarily in the central region with a cloud of gas and dust surrounding them as depicted in this two-piece 3D plate created from intensity of light and topographical information from the Hubble data.

The 3D print uses a combination of techniques to depict the stars, filaments, gas, and dust seen in optical light images from Hubble by applying special textures for each feature type and a height map that corresponds to the brightness observed in the data. The brightest feature captured in the data is the cluster of stars right in the core of NGC 602, represented by tall open circles.

WESTERLUND 2

Westerlund 2 is a cluster of thousands of young stars that are estimated to be only one to two million years old. Located about 20,000 light-years away from Earth, observations from Hubble in optical light reveal thick clouds where the young stars are forming. Infrared and X-ray studies of this system are also useful, to help us peer through the thick clouds of dust and gas. Westerlund 2 contains stars with masses over eighty times that of our Sun. Dense streams of matter steadily ejected by two such massive stars, called stellar winds, collide and produce a great amount of X-ray emission.

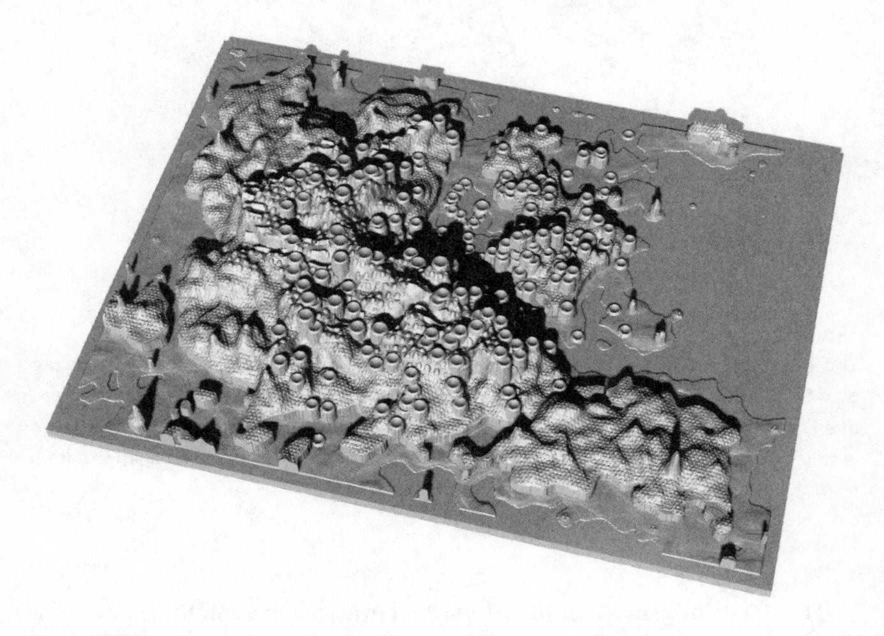

Figure 8.7
Located about 20,000 light-years from Earth, Westerlund 2 is a cluster of approximately three thousand young stars and astronomers estimate most are between one and two million years old. The intensity of the optical data along with its topographical features have been converted into this 3D plate showing the mass of young stars and thick dusty clouds, as well as a series of other filaments, pillars, ridges, and valleys.

Westerlund 2 is a particularly interesting star cluster in the Milky Way galaxy for scientists to study as it contains some of the hottest, brightest, and most massive stars known. The 3D plate of Westerlund 2 shows the mass of its young stars and thick dusty clouds, as well as other features such as filaments and pillars, as revealed in the visible light data.

CRAB NEBULA

In 1054 AD, Chinese sky watchers witnessed the sudden appearance of a "new star" in the heavens. Around the same time on the other side of the globe, Native Americans recorded the mysterious appearance of a similar

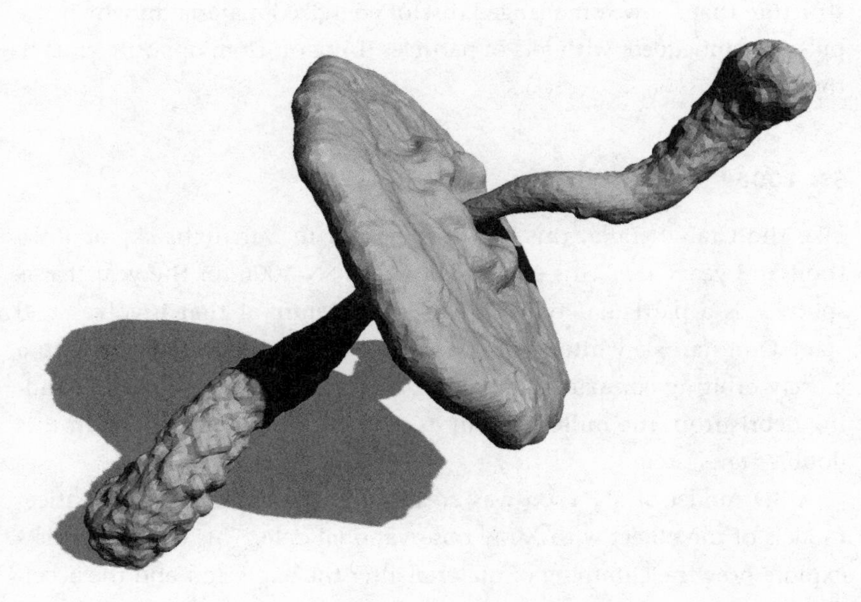

Figure 8.8
The Crab Nebula contains the remains of an exploded star being energized by a pulsar, which is sending out bursts of radiation thirty times a second. This model shows the inner X-ray structure of the Crab with a striking ringed disk and jets of particles firing off from opposite ends of the object.

object in petroglyphs. At the time, it was six times brighter than Venus, making it the brightest observed stellar event in documented history. Today, we know that this was the explosion of a star about 6,500 light-years from Earth; its remains are called the Crab Nebula.

Astronomers combined X-ray, infrared, and visible light data to generate a 3D model of a rotating Crab Nebula. The 3D structures serve as scientifically informed approximations for visualizing the nebula. The nested structures show that the nebula is not a classic supernova remnant, as once commonly thought. Instead of having a blast wave and debris from the supernova that has been heated to millions of degrees, the Crab may belong to a class of objects where the system's inner region consists of lower-temperature gas that is heated up to thousands of degrees when charged particles are moving close to the speed of light through magnetic fields. The 3D printable part of this visualized 3D system is the X-ray structure that shows off a ringed disk of energized material in which the pulsar is embedded, with jets of particles firing off from opposite ends of the object.

SN 1006

Like the Crab Nebula, this object appeared in our night sky about a thousand years ago. This object, known as SN 1006 for the year it was spotted, is a particular type of supernova remnant that involves two stars. One star is a white dwarf, the dense core of a Sun-like star with a closely orbiting companion star. The object we see now is the expanding debris from the millennia-old explosion of the white dwarf in this double star system.

A 3D model of SN 1006 was created by constraining mathematical models of the object with X-ray observational data. This helps scientists explore how the clumping of material after the explosion and the acceleration of high-energy particles affect the structure of the remnant. A ball of fiery-looking stellar debris and heavy elements has been shot into the interstellar medium with speeds of tens of thousands of miles per hour. The material is heated up to temperatures of tens of millions of degrees and glows brightly in X-ray light.

Figure 8.9
The supernova known as SN 1006 appeared in Earth's night sky over a thousand years ago. This 3D model of SN 1006 shows the clumped, highly energetic material after the explosion. The whole remnant, as well as cutouts of the blast wave and ejecta, are available for 3D printing.

NEUTRINO-DRIVEN SUPERNOVA

In certain types of supernova explosions, neutrinos play a major role. Neutrinos are nearly massless particles that can be produced when a star collapses into a neutron star, a city-sized superdense stellar core made up of essentially nothing but tightly packed neutrons. In this kind of supernova, neutrinos would escape from the star, taking energy with them and causing the star to cool much more rapidly than it would otherwise.

Researchers seek to better understand the physics of these stellar explosions. This 3D model simulates this variety of remnant left behind by a supernova explosion where neutrinos are prominent. It shows the distribution of iron, an important element that helps scientists track certain

Figure 8.10
Astronomers are trying to determine the role that neutrinos (particles with nearly no mass) play in the explosion of certain stars. The 3D model shows the irregular surface and lumpy material as well as the distribution of the iron, the forward and reverse shock waves detectable in certain kinds of light, and the line of sight to Earth.

physical processes. The timeframe for this model is about 350 years after the explosion, which is approximately how long it has been since the Cassiopeia A supernova remnant formed when it would have first been visible from here on Earth.

9

GOING BEYOND: GALAXIES, BLACK HOLES, AND MORE

We have established that stars are as glorious as they are plentiful, and that our Sun is just one of these billions of stellar gems. The sheer number of stars can be hard for the human brain to wrap itself around. Yet, there is much more to take in. There are billions of stars in just our Galaxy, which we call the Milky Way. There are equally vast amounts of stars in other galaxies, of which there are billions.

Galaxies are what the famed astronomer Edwin Hubble once called "island universes." The journey from early twentieth-century knowledge to our modern understanding is quite literally enormous. Before Hubble demonstrated the distance between these "spiral nebulas" and Earth, some leading astronomers of the day thought that our Galaxy encompassed the whole of the Universe.

What a difference fewer than a hundred years can make. Modern astronomy recognizes there are a range of types of galaxies that stretch across almost all cosmic time. Galaxies first show up in the Universe about a billion or so years after the Big Bang, or roughly thirteen billion years ago. (Some evidence suggests the seeds for the first galaxies may have started even before this.)

There are three main categories of galaxies today defined by shape: spiral, elliptical, and irregular. Our Milky Way is a spiral galaxy, which has a distinct set of spiral arms arranged in a relatively flat disk surrounding a large concentration of stars in its center known as the "bulge."

Of course, astronomers have had to determine this structure through many different lines of investigation because we are firmly embedded inside one of the spiral arms. Our technology is far from allowing us to glimpse a view of the Milky Way from an outside perspective. We can only view it from our vantage point on one of the Galaxy's spiral arms. We can, however, learn more about our Galaxy by studying analogs elsewhere. Luckily, spiral galaxies can be found in various orientations to Earth, giving us the means to study galactic doppelgangers and hence learn about our own.

Ellipticals are collections of older stars that, as the name suggests, are roughly oval in form. These galaxies often look less organized and have no distinct features such as spiral arms or bulges. Meanwhile, irregular galaxies are those that do not fit into either the spiral or elliptical category. The appearance of these galaxies, sometimes invoking thoughts of ink blots or abstract art, are byproducts of their past, which likely includes mergers or interactions with other galaxies.

One thing that galaxies do, in fact, have in common is the presence of black holes. These come in different sizes. The two main confirmed categories of these extreme objects are "stellar-mass" and "supermassive." Astronomers think stellar-mass black holes are formed when a particularly large star collapses onto itself and forms a gravitational warp in space. Supermassive black holes, in contrast, come from either a number of smaller black holes that merge over many millions of years, or a gigantic cloud of gas and dust that collapses onto itself.

The jury is still out on how black holes are formed, but there is a scientific consensus that supermassive black holes are found at the heart of most galaxies. Despite their reputation for destruction, black holes also act as galactic nurturers. They can regulate the growth of galaxies and contribute to the formation of stars. In other words, black holes and galaxies have a symbiotic relationship that appears to stretch to practically the very dawn of the Universe.

While black holes have no actual surface (they instead have a region, or boundary, called the event horizon that nothing can escape), 3D techniques have been a key in helping understand them. Because, as noted, black holes are active members of their community, by studying the regions around the black hole we can learn about not only them but also the objects in their vicinity. Black holes are the unseen drivers of many processes in the Universe—ranging from the formation of stars to the evolution of galaxies—and their influence happens beyond just two dimensions.

In recent years, a new line of exploration has opened to help scientists understand black holes and other violent and distant happenings across the Universe: gravitational waves. Predicted by Albert Einstein, these are permutations in the fabric of space-time itself. Using exceptionally sensitive lasers on the ground, scientists can measure the tiny distortions that pass through Earth when something powerful happens, such as two colliding black holes or a neutron star merger. Gravitational wave observations are well suited to 3D modeling and printing, allowing people to feel with their fingers the ripples created from a cataclysmic event billions of miles away.

And astronomers are not content with just knowing about what is in the Universe, they want to know about the Universe itself. The field of cosmology tries to answer questions like: how did the Universe begin? How old is it? What will happen in the future? By taking observations and melding these with theory, scientists have developed answers to some of our significant questions concerning the cosmos. These data, sometimes available across the entirety of the sky, can be explored through 3D modeling and 3D printing. This allows us in some instances to hold a depiction of the whole Universe in our hands.

Many of the objects in this chapter represent a particular method of bringing astronomical data into the 3D realm. For galaxies used in this type of cosmic cartography, the intensity of light detected by the various telescopes is translated into a tactile map that can be printed. In this type of 3D treatment, the height of peaks on the printed 3D object correspond to the intensity of light the telescopes collected.

SAGITTARIUS A*

Like most other galaxies, our Milky Way has a supermassive black hole at its center. Many astronomers, including those awarded the 2020 Nobel Prize for Physics, have been studying this object and the region around

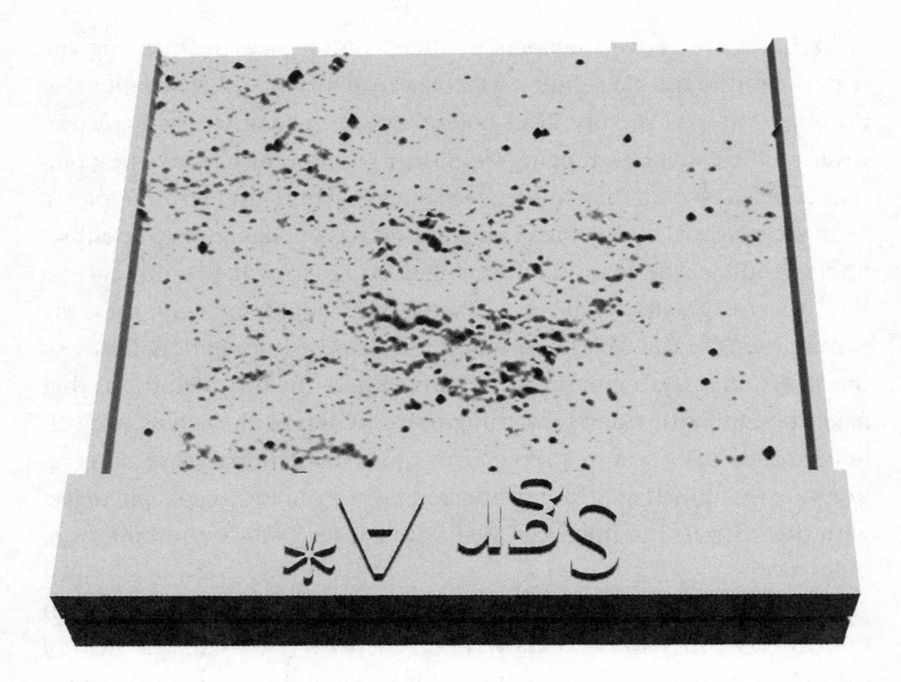

Figure 9.1
The closest supermassive black hole to Earth, Sagittarius A*, resides in the heart of our Milky Way galaxy. This 3D plate displays the intensity of the X-ray light and topographical features, with the black hole tucked away in the bright central region.

it for decades. Scientists estimate that this black hole, called Sagittarius A* (Sgr A*), contains the mass of four million Suns packed into a region smaller than our Solar System. As the closest supermassive black hole to Earth, Sgr A* has been extensively observed and studied using telescopes on the ground and in space. The Event Horizon Telescope project, which released the first image of the shadow of the black hole in the M87 galaxy in 2019, has also collected data on Sgr A*.

There is a great deal of gas and dust along the line of sight between Earth and Sgr A*, which is some 26,000 light-years away. Because of this, many kinds of light are absorbed. Certain kinds of light including X-rays, however, can penetrate this dusty veil and provide information about the region around Sgr A*. This 3D printable file is based on X-ray data from the Chandra X-ray Observatory and shows plumes around Sgr A* in the center that may be the remains of outbursts from the black hole that occurred millions of years ago. This 3D model covers an area that is about 90 light-years across, and the area of Sgr A* is marked with a plus or cross sign.

MILKY WAY

While Sgr A* is the heart of the Milky Way, it is by no means its only interesting part. Our Galaxy is a spiral, with majestic arms extending around the core known as the "bulge." (The Earth sits in one of its spiral arms, about two-thirds toward the edge.) The Milky Way is home to billions of stars and thousands of known planets, with countless more likely waiting to be discovered. The outer edges of the Milky Way are populated with ancient collections of stars known as globular clusters. The Milky Way also has an unseen, but detected, reservoir of dark matter, the mysterious substance that makes up most of the Universe.

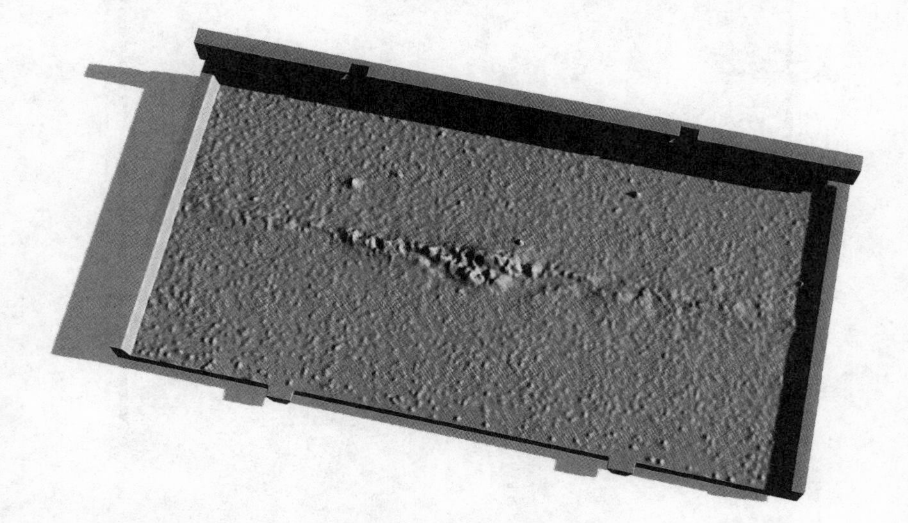

Figure 9.2

Our own Galaxy is called the Milky Way, and we live almost two-thirds of the way out on one of its beautiful spiral arms (called the Orion-Cygnus arm). Besides these arms, the Milky Way also features a core, or bulge, and is made up of billions of stars and planets as well as gas and dust. This 3D printable plate of the intensity map shows the Milky Way edge-on, toward its central region.

This 3D printable plate of our Milky Way provides a 360-degree panoramic view. We can explore the plane of the Milky Way as seen edge-on from Earth's vantage point, and view the flattened disk and the central bulge as well as numerous bright young stars, and even satellite galaxies.

M51

The Whirlpool Galaxy, officially known as Messier 51 or M51, is a majestic spiral galaxy that is oriented face-on to Earth as shown in this 3D printable plate. This gives us a spectacular view of a galaxy like our own that we cannot obtain of the Milky Way. M51 is not alone in space, rather

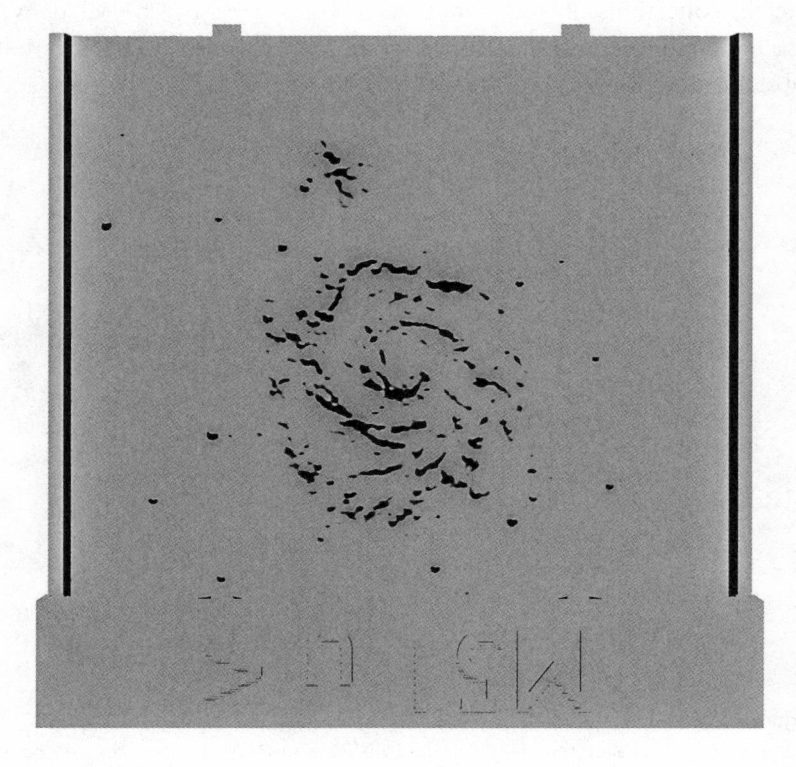

Figure 9.3
M51 is also known as the Whirlpool Galaxy because of its dramatic spiral shape. Being oriented face-on to Earth provides us with a clear view of the stars—both young and old—as well as gas and dust in this bright spiral galaxy.

TRAVELING TO OUR GALACTIC CENTER

To further explore the center of the Milky Way in 3D, you can venture about 26,000 light-years to the very center of our Galaxy in virtual reality. This immersive and interactive 3D experience, first released in 2018 and updated in subsequent years, enables the user to situate themselves in supercomputer simulations constrained by observations of the central three light-years around Sagittarius A* (Sgr A*), our Galaxy's supermassive black hole. Making the vantage point from the black hole, the visualization allows the user to pan through the complex structure, interacting with colliding wind from twenty-five massive stars, much of which is heated by shock waves—akin to sonic booms from supersonic aircraft—from eruptions from the black hole that are aglow in X-ray light.

it brushes a nearby small galaxy called NGC 5195. M51 is located about 31 million miles from Earth and can be spotted using a small telescope.

The spiral arms in M51 are easily the most prominent—and arguably beautiful—features of the galaxy to explore in this 3D printable version. They are also home to bursts of star formation, which may be enhanced with the tidal interactions M51 is experiencing from NGC 5195. Visible light telescopes reveal dust throughout the spiral arms that provide raw material for new stars to form, while X-ray observatories see the remains of supernovas and double systems containing stars paired with neutron stars or stellar-mass black holes. This is evidence that stars are forming at a furious rate, including young and massive ones that have short lives before undergoing a violent end.

M100

Astronomers sometimes refer to galaxies like M100 (and M51) as "grand design spirals." This lofty-sounding name is meant to highlight that the spiral arms are well defined, a difference from other types of spiral galaxies that have patchy or disrupted arms. Also known as NGC 4321, M100 is part of the Virgo galaxy cluster and is located about 54 million light-years from Earth. M100 is also interesting because it contains one of the youngest supernovas ever detected. First spotted by an amateur astronomer in 1979, the object known as SN 1979C (due to its discovery year)

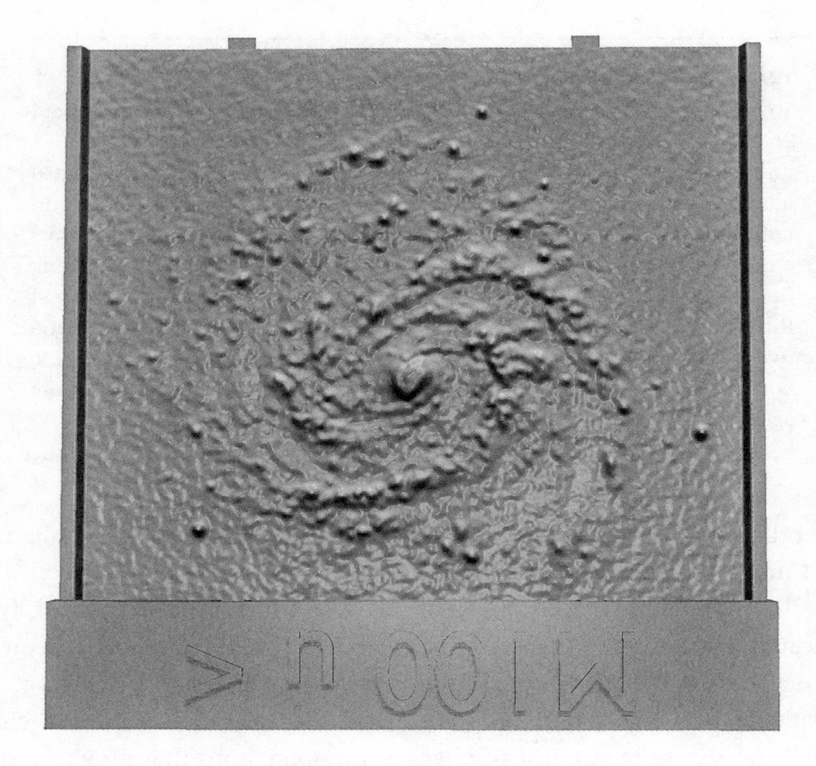

Figure 9.4
Messier 100 (M100), stretching about 160,000 light-years from one end to the other, is another spectacular spiral galaxy that is oriented face-on to Earth. In the 3D plate of M100 can be seen two prominent curved lanes of young stars as well as a population of older stars around its central region.

was found to be a bright source by orbiting X-ray telescopes. Evidence in the X-ray data indicates this object was a star about twenty times more massive than the Sun before it collapsed.

This galaxy has two prominent lanes of young stars as well as a population of older stars around its central region that can be explored in this 3D printable version. This galaxy stretches about 160,000 light-years from one end to the other, making it one and a half times the size of the Milky Way. There are two small galaxies nearby (not included in the 3D model), whose gravity may be influencing the shape of M100. In infrared light, astronomers see a prominent ring of hot, bright dust surrounding the inner core. Toward the outer edges of the galaxy where the spiral arms taper off, these infrared observations reveal large clumps of dust.

M86

All "M" galaxies get their name from Charles Messier, a French astronomer who died roughly around the time of Napoleon's reign as Emperor. Messier catalogued over a hundred objects that today we know include galaxies, star clusters, and nebulas. M86 is a galaxy that is generally classified a lenticular galaxy (a hybrid between an elliptical and a

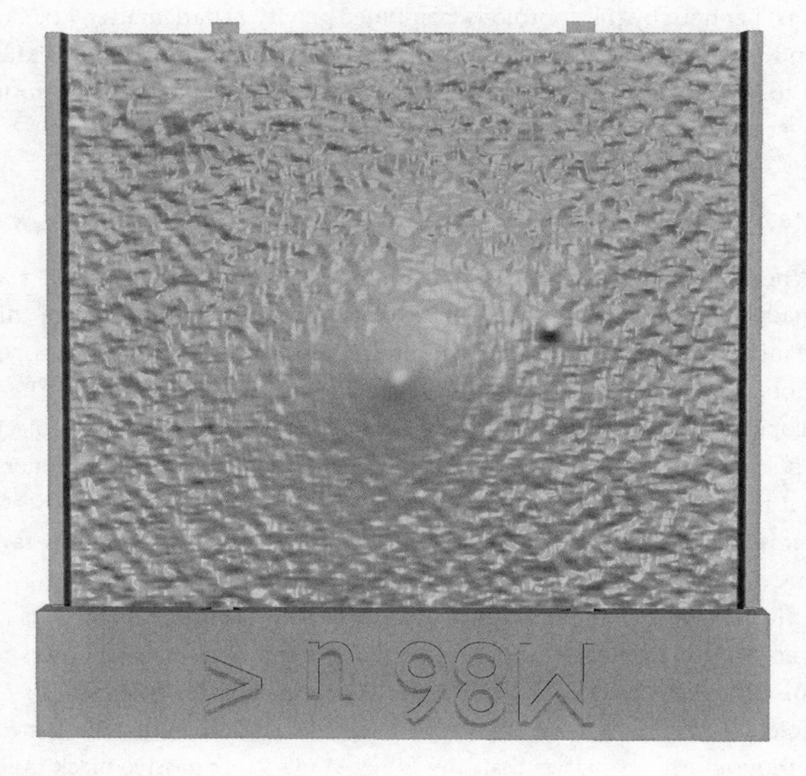

Figure 9.5
Messier 86 (M86) is an enormous lenticular galaxy (not quite an elliptical galaxy and not quite a spiral galaxy) that is part of the Virgo galaxy cluster. The 3D plate shows a bright central nucleus and includes globular clusters.

spiral). It is one of the brightest members of the Virgo galaxy cluster, a collection of 1,300 galaxies. Most of the galaxies in Virgo are moving away from Earth due to the expansion of the Universe, but astronomers have found that M86 is moving toward us. Since it sits about 60 million light-years from us, we do not have to worry about any close encounters with it for trillions and trillions of years (unlike the nearer Andromeda galaxy, which should merge with our Milky Way in five billion years or so).

Lenticular galaxies are mostly smooth and typically don't have prominent features like distinct arms or bulges. They can have a disk-like component, like those of spiral galaxies. The Virgo galaxy cluster has M86 accelerated to a high speed of some 1 million miles (1.6 million kilometers) per hour by the enormous, combined gravity of dark matter, hot gas, and hundreds of galaxies that comprise the cluster. The falling of M86 into Virgo is an example of how galaxy groups and galaxy clusters form over the course of billions of years.

M87

When the historic announcement of the first image of a black hole was made in April 2019, M87 instantly became famous. At the heart of this giant galaxy is a supermassive black hole that the array of radio telescopes around the globe trained their eyes on through the Event Horizon Telescope (EHT) project, which can be explored in this 3D printable object. This was not the first time M87 has gotten attention from astronomers. On the contrary, M87 has long been a target of telescopes that observe across the electromagnetic spectrum of light, from radio waves to X-rays and virtually every other type in between.

In addition to its photogenic black hole, M87 is home to trillions of stars and thousands of ancient globular clusters of stars. Like others in this 3D collection, M87 is in the Virgo cluster of galaxies. The giant black hole at its center has the mass equivalent to some 6.5 billion Suns, over a thousand times heftier than the Milky Way's supermassive black hole. This black hole is also powering an enormous jet that is moving in some places at 99 percent the speed of light when viewed through powerful X-ray telescopes, which we can see in the 3D model of those data.

Figure 9.6
M87 is the galaxy from which the first image of a black hole—or rather, the black hole's shadow or silhouette—was captured by the Event Horizon Telescope. The model shows the doughnut shape of light surrounding the supermassive black hole.

THE SOUTH POLE WALL

Not only do astronomers want to study galaxies, but they also want to understand how they fit together in larger structures. The South Pole Wall contains thousands of galaxies and stretches for some 700 million light-years. Unlike some of the other large-scale structures, the South Pole Wall is relatively close to the Milky Way. While astronomers have measured galaxies out to some 13 billion light-years, the farthest galaxy in the South Pole Wall is "only" around 600 million light-years away. It went unseen before because it is located behind the dust and gas of the plane of the Milky Way in our planet's southern sky.

Figure 9.7
The South Pole Wall is a tremendously large structure that contains thousands of galaxies and extends for about 700 million light-years. It is relatively close to the Milky Way at a distance of about 600 million light-years.

The South Pole Wall joins its cosmological cousins (the Great Wall, Boötes void, Sloan Great Wall, and others) in demonstrating that the Universe is not uniform—at least across certain scales. Due to their gravitational pull, galaxies often are drawn into groups and clusters, the latter of which can contain thousands of individual galaxies. These clusters can coalesce into long filaments, or chains, of galaxies, leaving large stretches of space empty. Astronomers sometimes refer to these tendrils of galaxies as comprising a cosmic web due to their shape. The Milky Way is part of a small cluster of galaxies known as the "Local Group," which sits on the edge of the Virgo cluster to which several galaxies in this chapter belong.

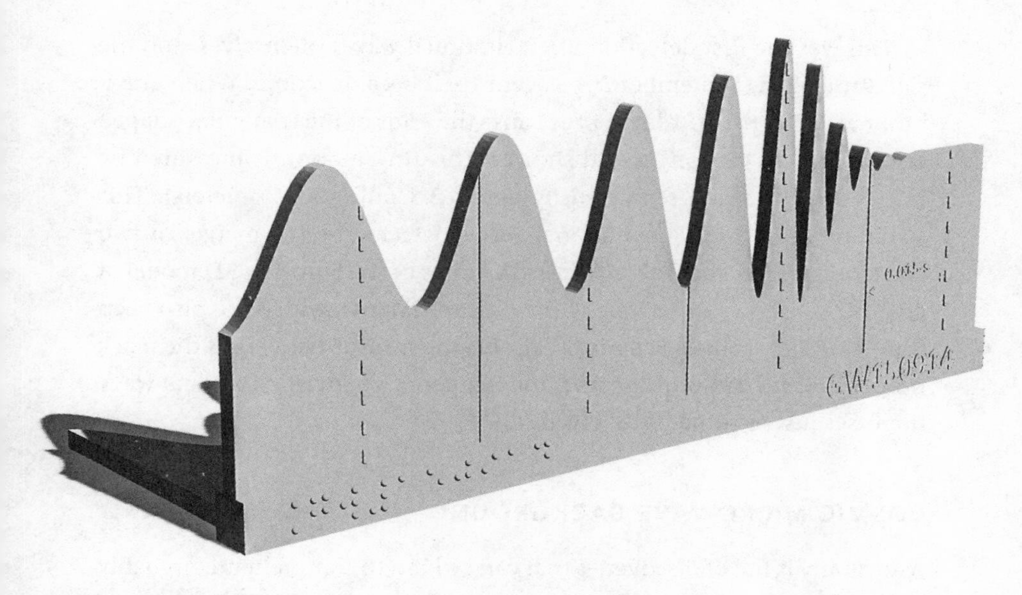

Figure 9.8
In 2015, astronomers detected gravitational waves from the merger of two black holes for the first time. The source became known as GW150914. The 3D model depicts the shape of the gravitational wave signal for the final 0.2 seconds of the data capture.

GW150914

On September 14, 2015, the Laser Interferometer Gravitational-Wave Observatory, better known as LIGO, detected signals at its two locations in Washington state and Louisiana. It took several months of careful checking, analysis, and preparation of scientific papers before the extremely large research team was ready to make its announcement. They did so on February 11, 2016, with a press conference that ushered in a new era of gravitational wave astronomy. Scientists must find these extremely tiny ripples amid a host of competing vibrations here on Earth, including everything from movements in the Earth's crust to heavy trucks going down a highway. A trio of scientists was awarded the 2017 Nobel Prize in Physics for their work on theories surrounding gravitational waves and their decades-long dedication to LIGO before success was achieved.

This was the first detection of gravitational waves themselves, and the signature of the September 2015 event itself became iconic. When gravitational waves pass through, they cause the arms of the triangular-shaped observatory to get longer and shorter. This tiny motion is measured by laser beams that are separated by about 2.5 miles (4 kilometers). This signal represents the gravitational waves detected by the merger of two black holes, each about thirty times the mass of the Sun. The 3D model of GW150914 shows this shape of the observed signal, which has also been translated into sound. Scientists call this the "chirp" because as the black holes spiral toward one another, the frequency of the gravitational wave increases just as some birds' chirps do.

COSMIC MICROWAVE BACKGROUND

Astronomy is full of discoveries that can be hard to comprehend. Arguably one of the biggest achievements yet has been measuring the imprint of the Big Bang. The European Space Agency's Planck telescope, which collected data from 2009 to 2013, measured tiny variations of temperature

PROBING DARK MATTER IN 3D

In the 1960s, astronomers Vera Rubin and her colleagues observed how quickly galaxies rotated. Surprisingly, she discovered that the acceleration of clouds orbiting the outer edges of spiral galaxies did not behave as they expected. Rather, there seemed to be at least several times the amount of matter, based on the motions of the stars throughout the spiral galaxies they studied, that was unaccounted for.

Over half a century later, scientists are still on the hunt to identify what this "dark matter" is, which constitutes most of the matter in the Universe today. One way that astronomers are tackling this problem is by using 3D techniques to map where dark matter exists. Using a phenomenon called "gravitational lensing," astronomers can look for this hidden matter that pervades the Universe and use the information to map where it falls across the expanse of space. First proposed by Albert Einstein, gravitational lensing involves the warping of space-time and the bending of light from distant objects. The amount of distortion that astronomers see from these far-flung galaxies and galaxy clusters allows them to discern how much unseen dark matter exists over vast swaths of the Universe.

Figure 9.9
Also known as the "imprint" of the Big Bang, the cosmic microwave background model shows the various dimples and bumps that would eventually form into the Universe we know today. This model, when 3D printed, allows you to hold the whole, albeit very early, Universe in your hand.

across the sky that have helped scientists pin down the age of the Universe (13.8 billion years old) as well as important features such as its shape and composition.

This 3D model shows the cosmic microwave background, or CMB, measured by Planck. Each of these tiny dimples and bumps represents minute differences in temperature and density. About four hundred thousand years after the Big Bang, the Universe had cooled enough for simple atoms to form. Over the billions of years since, these small perturbations in the CMB evolved into the structures around the Universe we see today including galaxies, galaxy clusters, and the mega-structures like the South Pole Wall.

10

LOOKING FORWARD (AND THROUGH) IN THE 3D UNIVERSE

We hope that this book has provided a taste (and, perhaps, a touch) of how far we have come in understanding our complex Universe and how 3D technology is playing a vital role. But these remarkable achievements also lead to another question: where do we go from here? The future of 3D modeling and printing in astronomy is certainly bright.

In some ways, we can think of ourselves as entering an era of Big Data astronomy. Already, scientists have collected vast amounts of data across the entire sky and in a range of types of light. Successfully mining the already available data has led to extraordinary discoveries, and an avalanche of forthcoming data will undoubtedly lead to countless more.

It's astonishing to think that in the past decade we have developed hardware and software that allow us to hold versions of real data of actual cosmic objects. Light collected by magnificent telescopes after traveling for millions or billions or even more miles and kilometers can end up being held in the hands of senators, school children, scientists, artists—virtually any of us. Not only that, but we can already walk around inside these objects in space in virtual reality and augmented reality and interact with them through holograms.

As is frequently the case in science, new advances lead to more questions. We ask ourselves: what comes next? What is the potential impact of 3D modeling our Universe and the objects within it? What will we learn when we can more strategically assess the greater Universe around us by working in three dimensions from the start? That is, what happens when we think about the possibilities in 3D at the very beginning of collecting data, instead of tacking it on at the end of the pipeline as an extra benefit of the data?

There are new telescopes being constructed now, and even more new telescopes planned that will have many times the collecting area, the resolution, and the overall vision capabilities of those now in use. These planned projects in the coming years and decades will capture stupendous amounts of higher-resolution data, more velocity and depth information, and at even greater distances and larger scales than scientists have yet been able to achieve. There will be a veritable sandbox of data, albeit of enormous digital dimensions, to play in. Astronomers, computer or data scientists, makers, creators, and researchers in related fields need to think about these possibilities today if we are going to be able to make new strides in our understanding of the 3D Universe tomorrow and in the years to come.

Meanwhile, technologists are not sitting idly by. Augmented reality glasses are on the horizon and virtual reality experiences are being greatly improved with touch, sound, and multiplayer technologies. People working in separate locations—either down the hall or across the globe—are gaining the ability to use hand movements to control and manipulate data together in extended reality in real time. Where will such multimodal and collaborative technology take us in our exploration of the Universe?

This brings us back to where we started: making a connection to the cosmos. The truth is we are part of the Universe, and it is quite literally part of us. Whether it is the calcium in our bones or the iron in blood, "we are star stuff" as Carl Sagan famously said. Our chemical heritage can be scientifically traced to previous generations of stars, with the life cycles of stars providing the elements necessary for life as we know it.

Beyond honoring that indisputable link, our place in the cosmos can provide some well-needed perspective. Certainly, a distant galaxy does not care for our grievances with one another (of course, it does not feel anything at all). We share this one rocky planet around this middle-aged, average-sized star in a nondescript arm of an ordinary galaxy. We may have our differences on this world, but a quick look around space and we notice that we may want to acknowledge our commonality or recognize it is a lonely existence.

Exploring the Universe in three dimensions is not a magical panacea to all that ails us as a species. But maybe it can serve as an entry point for some people to begin to explore our common cosmos. For others, it could be a chance to delve deeper. And maybe for yet another audience, it is an innovation on the cutting edge of science and technology and that is what appeals to them.

Regardless of the motivation, the 3D Universe is ours to discover. While we certainly do not know exactly what the future holds, we think that the three-dimensional exploration of space will continue to increase access to and provide information about the wonders that humanity has discovered. It is indeed a marvelous Universe that we inhabit, and 3D gives us an entry pass that is different from anything we have ever had before.

ACKNOWLEDGMENTS

No book would be complete without us expressing our deepest gratitude to our families and long-time collaborators. We would like to graciously thank Peter Edmonds, who is both our friend and colleague, for his scientific expertise.

Many thanks to Jackson Arcand, research assistant and visualization specialist extraordinaire for this book who was particularly helpful while Kim dealt with some untimely carpal tunnel syndrome effects. And as always, Kim is deeply grateful to her family for their unwavering love and support, particularly her husband John, daughter Clara, and her mother Chris. No thanks go to her dog Juno and cat Midnight who were very distracting and constantly begging for treats!

Megan could never complete a project like this without the love, patience, and support of her family. Her wife Kristin and four kids (Anders, Jorja, Iver, and Stella) give her the time away from the daily demands of everything here on the ground so she can let her thoughts wander up into space.

GLOSSARY

3D print — Creation of a physical copy of a three-dimensional model, typically by a process of additive manufacturing on demand, in which a type of material, for example, plastic, metal, or sugar, is continually added layer by layer to formulate the object.

Augmented reality (AR) — A virtual environment that provides the user with an enhanced version of reality, for example, when digital materials are added over a typical view of a person, place, or thing. Pokémon Go and Snapchat filters are two examples of popular AR-enhanced applications.

Big Bang — The event that most astronomers consider to be the beginning of the universe, in which space-time originated in a state of enormously high temperature and density and subsequently expanded and cooled.

Blueshift — Motion-induced change in the observed wavelength from a source that is moving toward us (see "redshift").

Binary code A system that uses two digits (1 and 0) to represent information, like an "on" and "off" position of a switch.

Black hole A dense, compact object whose gravitational pull is so strong that—within a certain distance of it—nothing can escape, not even light. Black holes are thought to result from the collapse of certain very massive stars at the end of their evolution.

Cosmic microwave background (CMB) The microwave radiation coming from all directions that scientists think is the glow left over from the Big Bang.

Cosmology The study of the origin and evolution of the Universe as a whole.

Dark matter A mysterious and invisible substance that constitutes most matter in the Universe.

Doppler shift (or Doppler effect) Apparent change in wavelength of the radiation from a source due to its relative motion away from or toward the observer.

Electromagnetic spectrum The full range of light, including radio waves, infrared light, microwaves, optical light, ultraviolet radiation, X-rays, and gamma rays.

Elliptical galaxy Category of galaxy in which the stars are distributed in an elliptical shape on the sky, ranging from highly elongated to nearly circular in appearance.

Event horizon Imaginary spherical surface surrounding a black hole within which no event can be seen, heard, or known about by an outside observer.

Extended reality (XR) Umbrella term that describes the combined category of virtual reality, augmented reality, and mixed reality environments.

Frequency The number of wave crests passing any given point in a given period of time.

Galaxy A system of planets, stars, gas, dust, and dark matter held together by gravity.

Galaxy cluster A collection of hundreds or thousands of individual galaxies immersed in a huge reservoir of hot gas and dark matter.

Gravity The attractive effect that any massive object has on all other massive objects. The greater the mass of the object, the stronger is its gravitational pull.

Gravitational wave The gravitational analog of an electromagnetic wave whereby gravitational radiation is emitted at the speed of light from any mass that undergoes rapid acceleration.

Hubble Space Telescope The first large optical telescope launched above the Earth's atmosphere (in April 1990) and carrying instruments sensitive to visible and ultraviolet light.

Light-year The distance that light, moving at a constant speed of 650 million mph or 300,000 km/s, travels in one year. One light-year is about 6 trillion miles (10 trillion kilometers).

Magnetic field Field that accompanies any electric current or changing electric field and governs the influence of magnetized objects on one another.

Mass A measure of the total amount of matter contained within an object.

Milky Way galaxy The spiral galaxy to which the Sun and Earth belong.

NASA National Aeronautics and Space Administration, the United States agency for space exploration.

Nebula An interstellar cloud of gas and dust.

Neutrino An electrically neutral elementary particle that has little or no mass, moves at close to the speed of light, and interacts very weakly with matter.

Neutron star A dense stellar remnant produced by the collapse of the core of a massive star as part of a supernova that destroys the rest of the star.

Pulsar A rotating neutron star that emits pulses of light.

Quasar A very bright object that is powered by material falling onto a supermassive black hole in the center of a galaxy.

Red giant star An evolved star that has exhausted the hydrogen fuel in its core and is powered by nuclear reactions in a hot shell around the stellar core.

Spectral line A feature observed in emission or absorption at a specific frequency or wavelength.

Spiral galaxy Galaxy composed of a flattened, star-forming disk component that may have spiral arms and a large central galactic bulge.

Stellar evolution The changes experienced by stars as they are born, mature, and grow old.

STEM An acronym for "science, technology, engineering, and math."

Supernova remnant The expanding debris field left over from a star that exploded.

T Tauri A class of very young, often flaring stars.

Ultraviolet Type of light just outside the visible range, corresponding to wavelengths slightly shorter than blue light.

Virtual reality (VR) A technology that simulates a participant's physical self in computer-generated environments that adjust to the user's presence to provide a sense of immersion.

Visible light The small range of the electromagnetic spectrum that human eyes perceive as light.

White dwarf A star that has exhausted most or all its nuclear fuel and has collapsed to a very small size; such a star is near its final stage of life.

Wolf-Rayet Hot massive stars in a late stage of their evolution that are losing mass through winds.

X-ray Region of the electromagnetic spectrum corresponding to radiation of high frequency and short wavelengths, far outside the visible spectrum.

FIGURE CREDITS

Chapter 2: Figure 2.1: NASA/CXC

Chapter 3: Figure 3.1: ESA/Gaia/DPAC, CC BY-SA 3.0 IGO, https://creativecommons.org/licenses/by-sa/3.0/igo; figure 3.2: Alyssa A. Goodman, Erik W. Rosolowsky, Michelle A. Borkin, Jonathan B. Foster, Michael Halle, Jens Kauffmann, and Jaime E. Pineda; figure 3.3: ESO

Chapter 4: Figure 4.1: NASA/Emmett Given

Chapter 5: Figures 5.1–5.6: Star Coin Project by Bruce Bream, CC BY-NC-SA 4.0, https://creativecommons.org/licenses/by-nc-sa/4.0/

Chapter 6: Figures 6.1 and 6.2: Michael Carbajal, NASA Headquarters; figure 6.3: Brian Kumanchik and Christian Lopez, NASA/Jet Propulsion Lab (JPL)-Caltech; figure 6.4: NASA Johnson Space Center; figure 6.5: Digital Corporation; figures 6.6 and 6.7: Brian Kumanchik and Christian Lopez, NASA/JPL-Caltech; figure 6.8: NASA; John Doroshenko, Rutgers University; figures 6.9–6.11: Brian Kumanchik and Christian Lopez, NASA/JPL-Caltech; figure 6.12: NASA/VTAD; figure 6.13: NASA Goddard Space Flight Center

Chapter 7: Figures 7.1 and 7.2: Doug Ellison, NASA/JPL; figure 7.3: Richard Kim; figure 7.4: NASA/AMES; figure 7.5: Doug Ellison, NASA/JPL; K. Gwinner, J. Oberst, R. Jaumann, and G. Neukum, ESA/DLR/FU Berlin; figure 7.6: Doug Ellison, NASA/JPL; figure 7.7: Kris Capraro, NASA/JPL; figures 7.8 and 7.9: Doug Ellison, NASA/JPL; figure 7.10: Michael C. Nolan, Arecibo Observatory/NASA/NSF; figure 7.11: Doug Ellison, NASA/JPL; figure 7.12: Bougnatophile, CC BY-SA 4.0, https://creativecommons.org/licenses/by-sa/4.0/deed.ast; figure 7.13: Francis Reddy, University of Maryland, College Park, and NASA Goddard Space Flight Center

Chapter 8: Figure 8.0: Melissa Weiss, NASA/CXC; figure 8.1: Salvatore Orlando, INAF-Osservatorio Astronomico di Palermo; figure 8.2: W. Steffen, M. Teodoro, T. I. Madura, J. H. Groh, T. R. Gull, A. Mehner, M. F. Corcoran, A. Damineli, and K. Hamaguchi, ESO/Hubble/VLT; figure 8.3: A. F. McLeod (ESO, Garching, Germany), J. E. Dale (Universitäts-Sternwarte München, München, Germany; Excellence Cluster Universe, Garching bei München, Germany), A. Ginsburg (ESO), B. Ercolano (Universitats-Sternwarte Münchene; Excellence Cluster Universe), M. Gritschneder (Universitats-Sternwarte München), S. Ramsay (ESO), and L. Testi (ESO; INAF/Osservatorio Astrofisico di Arcetri, Firenze, Italy), ESO/Hubble; figure 8.4: Anne Archibald, Anton Pannekoek Institute; figure 8.5: Tom Hendrix, Rony Keppens, Allard Jan van Marle, Peter Camps, Maarten Baes, and Zakaria Meliani; figures 8.6 and 8.7: Carol Christian (STScI), Antonella Nota (ESA and STScI), and Perry Greenfield (STScI), NASA/STScI; figure 8.8: NASA, ESA, F. Summers, J. Olmsted, L. Hustak, J. DePasquale, G. Bacon (STScI), N. Wolk (CfA/CXC), R. Hurt (Caltech/IPAC); figures 8.9 and 8.10: Salvatore Orlando, INAF-Osservatorio Astronomico di Palermo

Chapter 9: Figure 9.1: NASA/CXC/SAO; Nicolas J. Bonne, Tactile Universe; figures 9.2–9.6: Nicolas J. Bonne, Jennifer A. Gupta, Coleman M. Krawczyk, and Karen L. Masters, "Tactile Universe Makes Outreach Feel Good, *Astronomy & Geophysics* 59, no. 1 (February 2018): 30–33, https://doi.org/10.1093/astrogeo/aty028; figure 9.7: Virgo Outreach team; figure 9.8: Daniel Pomarède Institut de Recherche sur les Lois Fondamentales de l'Univers, CEA Université Paris-Saclay; figure 9.9: Dave Clements, Suzu Sato, and Ana Portela Fonseca (Imperial College London), ESA/Planck

Color insert: Plate 1: NASA; plate 2: J. Vaughan, NASA/CXC; plate 3: NASA/JPL-Caltech; plate 4: NASA; plate 5: NASA/JPL-Caltech; plate 6: NASA/JPL-Caltech/MSSS; plate 7: NASA/JPL; plates 8 and 9: NASA; plates 10 and 11: NASA/GSFC/Arizona State University (ASU); plate 12: NASA; plate 13: NASA/GSFC/ASU; plate 14: NASA/JPL-Caltech; plate 15: NASA/JPL/University of Arizona; plate 16: NASA/JPL-Caltech; plate 17: NASA/JPL-Caltech/UCLA/MPS/DLR/IDA; plate 18: NASA/GSFC/University of Arizona; plate 19: NASA/DSWCO/EPIC; plate 20: NASA/GSFC; plate 21: N. Smith (University of Arizona) and J. Morse (BoldlyGo Institute), NASA, ESA; plate 22: NASA/STScI; plate 23: Adam Block, Mount Lemmon SkyCenter, University of Arizona, CC BY-SA 3.0, https://creativecommons.org/licenses/by-sa/3.0/; plate 24: NASA, ESA, the Hubble Heritage Team (STScI/AURA), A. Nota (ESA/STScI), and the Westerlund 2 Science Team; plate 25: M. Robberto, R. C. O'Dell, L. A. Hillenbrand, M. Simon, D. Soderblom, E. Feigelson, J. Krist, P. McCullough, M. Meyer, R. Makidon, J. Najita, N. Panagia, F. Palla, M. Romaniello, I. N. Reid, J. Stauffer, K. Stassun, K. Smith, B. Sherry, L. E. Bergeron, V. Kozhurina-Platais, M. McMaster, and E. Villaver, NASA/ESA/STScI; plate 26: NASA/STScI; plate 27: X-ray: NASA/CXC/SAO; Optical: NASA/STScI; Infrared: NASA-JPL-Caltech; plate 28: F. Winkler, NASA/CXC/Middlebury College; plate 29: Y. Bai, A. J. Barger, V. Barger, R. Lu, A. D. Peterson, and J. Salvado; NASA/CXC/University of Wisconsin; plate 30: Event Horizon Telescope Collaboration; plate 31: X-ray: D. Wang, NASA/CXC/UMass; Optical: D. Wang, NASA/ESA/STScI; IR: S. Stolovy, NASA/JPL-Caltech/SSC; plate 32: X-ray: R. Kilgard, NASA/CXC/Wesleyan University; Optical: NASA/STScI; plate 33: NASA, ESA, and Judy Schmidt; plate 34: Sloan Digital Sky Survey; plate 35: LIGO/Axel Mellinger; plate 36: NASA; plate 37: Wikipedia/pithecanthropus4152; plate 38: Wikipedia/Łukasz Mularczyk; plate 39: Wikipedia/Scott Roy Atwood; plate 40: Wikipedia/Roberto Mura

INDEX

Page numbers in italics refer to figures.